Design-for-Test and Test Optimization Techniques for TSV-based 3D Stacked ICs

Brandon Noia • Krishnendu Chakrabarty

Design-for-Test and Test Optimization Techniques for TSV-based 3D Stacked ICs

Foreword by Vishwani Agrawal

Springer

Brandon Noia
ECE, Duke University
Durham, NC, USA

Krishnendu Chakrabarty
ECE, Duke University
Durham, NC, USA

ISBN 978-3-319-02377-9 ISBN 978-3-319-02378-6 (eBook)
DOI 10.1007/978-3-319-02378-6
Springer Cham Heidelberg New York Dordrecht London

Library of Congress Control Number: 2013955016

Printed on acid-free paper

Springer is part of Springer Science+Business Media (www.springer.com)

To my parents for their boundless support.
—Brandon Noia

To all my truly outstanding students over the years.
—Krishnendu Chakrabarty

Foreword

Present trends in the electronics industry indicate that the three-dimensional stacked integrated circuit (3D SIC) is a technology with potential for wide scale application. So, is this 3D SIC just an extension of a flat (2D) integrated circuit? If this device is three dimensional then are we going to build 4D IC in the future as our world is believed to have a fourth dimension of time?

Let me settle the last question first. The use of the fourth dimension (time) is not new for digital circuits. Reuse of logic gates over time is allowed by storage elements or flip-flops, without which the implementation of a digital function, if at all possible, will require many more gates. I would argue that our 2D chips already have three dimensions, x, y, and t. We just have a habit of ignoring time while specifying the dimensions in a chip. Evidently, the 3D SIC has the fourth dimension.

Now, to address the first question let me start by examining the flat 2D IC. It contains a layer of active components and multiple levels of interconnects. The same is true for a printed circuit board (PCB) of the previous generation electronics. These are like a multi-story building in which the living space is limited to just one floor and all other floors only provide corridors connecting the rooms on that single floor. Sounds unnatural? But that is what we have been doing in 2D ICs with multiple routing levels and multi-layer PCBs. A successful design of a multi-story building has functional living space on all floors with corridors within floors and elevators between floors. Of course, each floor may be customized for a specific function such as family housing, office, storage, shopping, or restaurant. This concept of a multi-story building in which our society has a long experience is now being applied to build integrated circuits. Indeed, the benefits can be extrapolated from one domain to other.

However, as one would say, "It's no use building skyscrapers unless you can have working elevators." For a 3D SIC, a through silicon via (TSV) is the elevator. The IC manufacturing technology allows us to fabricate TSVs but to ensure that they work, we must test them and, if found broken, fix them.

TSV testing is one aspect in which a 3D SIC differs from a conventional flat or 2D IC. Besides, there is not just one way to realize a 3D SIC, not at this time anyway as this technology is still evolving. For example, it is found that stacking of wafers may be more economical than stacking of chips. Stages in fabrication give rise to various test scenarios. Just for TSVs, there may be a pre-bond test and multiple post-bond tests.

An IC engineer may be familiar with the prevailing methods and tools, but the emerging 3D SIC technology is an application domain that is unchartered. This book is a guide. Its nine chapters give a clear exposition of problems of testing and their solutions. Some topics are so new that the authors, Brandon Noia and Krishnendu Chakrabarty, must rely on their own research. They have excelled in this.

The book begins with outlines of the 3D SIC manufacturing process and its test challenges. Wafer matching with repositories is presented as a practical method. Examples of manufacturing flows suitable for 3D memory architectures are given. In this context, redundancy methods for improving yield and testing cost are shown to be beneficial. These topics constitute the first two chapters.

High TSV yield is essential for ensuring an economical stack assembly. This motivates pre-bond TSV test for which there are at least two methods, namely, built-in self-test (BIST) and probe test. Chapters 3–5 give details of these two alternatives.

In Chap. 6, the authors examine impacts of the design for testability hardware. A major impact is on performance. Because the test hardware surrounds TSVs that communicate signals between the stacked chips, a proper partitioning of the system can improve the performance. The authors present a 3D optimization flow utilizing retiming with logic decomposition and redistribution across dies. This is shown to reduce the impact of the test architecture on the functional timing of inter-die paths.

Chapters 7 and 8 discuss ideas on implementing the test hardware and test optimization. The P1838 die wrapper and JEDEC I/O emerging standards for 3D stacks are introduced. An integer linear program (ILP) formulation is proposed for optimally utilizing the available test access mechanism (TAM) and developing test schedules with reduced overall test time.

The book has eight major chapters with a final ninth chapter, titled Conclusion, informing the reader where to find what. There is an impressive list of references. I congratulate the authors Brandon and Krishnendu on the publication of this book. It is an important contribution to the development of the emerging technology of 3D stacked integrated circuits.

Auburn, AL, USA — Vishwani D. Agrawal

Preface

As integrated circuits (ICs) continue to scale to smaller dimensions, relatively long interconnects have become the dominant contributor to circuit latency and a significant component of power consumption. In order to reduce the length of these interconnects, 3D integration, and in particular 3D stacked ICs (3D SICs), have become a significant area of research in both academia and industry. 3D SICs not only have the potential to reduce average interconnect length and alleviate many of the problems caused by long global interconnects, but they can offer greater design flexibility over 2D ICs, significant reductions in power consumption and footprint in an era of mobile applications, increased on-chip data bandwidth through latency reduction, and improved heterogeneous integration.

Compared to 2D ICs, the manufacture and test of 3D ICs is significantly more complex. Through-silicon vias (TSVs), which function as dense vertical interconnects in a die stack, are a source of additional and unique defects not seen before in integrated circuits. At the same time, testing these TSVs, especially before die stacking, has been a significant challenge to academic and industrial test engineers. Testing a 3D stack is constrained by limited test access, test pin availability, power, and thermal constraints. Therefore, efficient and optimized test architectures are needed to ensure that pre-bond, partial, and complete stack testing is not prohibitively expensive.

This book is meant as a guide for industry designers, university professors, and students, either as a textbook for learning about 3D IC testing or as a complete view of the cutting-edge research being performed in the field. For the student, this book provides a thorough introduction to the benefits and challenges of 3D integration and the associated difficulties for 3D test, as well as insights into state-of-the-art in test solutions and test optimization. For the academic researcher, this book further explores the literature in 3D test, guiding the reader through the strengths and weaknesses of present solutions and test questions that remain unanswered. This book is the perfect jumping-off point for anyone looking to perform further research in the field. For the industry engineer, this book contains a wide array of state-of-the-art 3D test architectures and optimizations along with the detailed results and in-depth analysis required to best choose and integrate valuable concepts into real-world designs. Furthermore, this book examines and explains the emerging standards that will likely drive industry test integration for years to come.

This book broadly explores three important categories of 3D test—pre-bond test, post-bond test, and test optimization. The book begins by examining pre-stacking optimization and pre-bond test before moving to post-bond test and optimization. Test solutions, including BIST and probing for pre-bond test and emerging standards for post-bond test, are thoroughly covered, with additional insights and test optimizations explored in between.

At the start of the book, Chap. 1 provides a brief overview of 3D integration technology, common design-for-test features, and the unique test challenges introduced by 3D integration.

Chapter 2 discusses wafer matching and 3D memory test. This chapter explores the varieties of repositories, matching algorithms, matching criteria, and other important considerations for wafer matching and their impact on the yield and cost of 3D stacks. It also examines the differences between 2D and 3D fault models and test and repair architectures for memory, and explains state-of-the-art solutions available in the literature on memory test.

Chapter 3 discusses the benefits and drawbacks of built-in-self-test (BIST) for pre-bond TSV testing. A detailed analysis of the TSV pillar and associated defects is provided. The chapter explores a number of BIST architectures, including memory-like test, voltage division with repair, and ring-oscillators. It provides an in-depth look at the kinds of defects each technique can detect and the accuracy with which detection is achieved.

Chapter 4 covers the alternative solution to BIST for pre-bond TSV test—pre-bond TSV probing. An introduction to present probe-card technology for 2D test and future probe-card solutions for 3D test is provided. The bulk of the chapter focuses on a technique for probing multiple TSVs at a time that is compatible with present probe-card technologies. Detailed results and analysis are provided discussing the feasibility and accuracy of the method. An optimization method, with associated results, is provided to reduce pre-bond TSV test time by testing multiple TSVs simultaneously through a single probe needle.

A number of optimizations are then provided to further reduce the cost of pre-bond test. Chapter 5 presents a detailed examination of performing pre-bond scan test by reusing the test architecture presented in Chap. 4. The feasibility, speed, and overhead of performing pre-bond TSV and structural test under the same paradigm is fully explored. Chapter 6 then presents a retiming-based test architecture optimization for reducing the impact of the architectures of Chaps. 5 and 4 on the post-bond functional mode of the 3D stack.

Chapter 7 covers the emerging test standards targeted at 3D SICs. This includes die-level wrappers, which ensure that dies in a stack present a standardized interface for pre-bond test and post-bond integration. The chapter further examines the test features of the JEDEC standard being proposed for high-speed memory-on-logic stacks.

Chapter 8 covers an optimization technique for reducing the test time of a post-bond stack. The optimization takes into account 3D-specific test constraints such as dedicated test TSVs and test access only through the bottom die. Furthermore, it can optimize the test architecture and test schedule of a stack when performing any or all possible partial- and complete-stack tests.

Finally, Chap. 9 concludes this book with a look back on the topics covered and last thoughts by the authors.

Whether you are new to 3D test and looking to get your feet wet or a seasoned veteran, the authors hope that the scope and readability of this text will provide you with what you need to contribute and excel in the field of 3D test.

Durham, NC, USA

Brandon Noia
Krishnendu Chakrabarty

Acknowledgements

The authors acknowledge the financial support received from the National Science Foundation and the Semiconductor Research Corporation. The authors thank Erik Jan Marinnisen for a very fruitful collaboration over the years. The authors also thank Sun Kyu Lim and Shreepad Panth for the 3D design benchmarks that they contributed for use in research. Finally, the authors acknowledge the contributions of the past and present students at Duke University, including Mukesh Agrawal, Sergej Deutsch, Hongxia Fang, Sandeep Goel, Yan Luo, Fangming Ye, Mahmut Yilmaz, Zhaobo Zhang, and Yang Zhao.

Brandon Noia would like to thank his parents whose dedication, sacrifice, and example has been the greatest constant through the triumphs and failures of life. Good parenting can offset the many flaws of a person's character, and his parents are exceptional. He would like to thank his brother whose constant companionship lasted longer than it likely should have, but that meant all the difference through the years. Finally, he would like to thank his fiancée, whose boundless support and affection has been far more than he deserves.

Contents

Chapter 1
Introduction

The semiconductor industry has relentlessly pursued smaller device sizes and low-power chips in a broad range of market segments, ranging from servers to mobile devices. As transistors continue their miniaturization march through smaller technology nodes, the limits of device scaling tend to be reached. Interconnects, particularly global interconnects, are becoming a bottleneck in integrated circuit (IC) design. Since interconnects do not scale as well as transistors, long interconnects are beginning to dominate circuit delay and power consumption.

To overcome the challenges of scaling, the semiconductor industry has recently begun investigating 3D stacked ICs (3D SICs). By designing circuits with more than one active device layer, large 2D circuits can instead be created as 3D circuits with significantly shorter interconnects. 3D SICs will therefore lead to a reduction in the average interconnect length and help obviate the problems caused by long global interconnects [25, 28, 29]. This not only leads to large reductions in latency, but can also lead to lower-power, higher-bandwidth circuits with a higher packing density and smaller footprint. Since dies in a 3D stack can be manufactured separately, there are also benefits to the heterogeneous integration of different technologies into a single 3D stack.

This book assumes that the reader is familiar with basic concepts in circuit test. The reader should already be familiar with the purpose and variety of tests, especially structural test, with the concepts of fault models, automated test equipment (ATE), automated test pattern generation (ATPG), and many of the test standards and architectures used for testing 2D circuits. If the reader is unfamiliar with these concepts, or would like a reference to go along with this book, several relevant books will provide the needed material [1, 2]. This book is also not a primer on the manufacture and design of 3D monolithic or stacked ICs, though it will cover those topics related to 3D manufacturing that are important to a test engineer. For a more complete primer on 3D IC manufacturing, other books are recommended [3].

This book will instead cover a wide array of topics regarding 3D circuit test, from wafer matching to pre-bond BIST techniques, and from memory test to wafer probing. This introduction will serve as a quick overview of testing and the need

B. Noia and K. Chakrabarty, *Design-for-Test and Test Optimization Techniques for TSV-based 3D Stacked ICs*, DOI 10.1007/978-3-319-02378-6_1,

for and motivate the need for 3D DfT and test optimization. The remainder of this book provide in-depth description, results, theoretical advances, architectures, and optimization methods for 3D circuit test.

Chapter 2 discusses two related topics—wafer matching and 3D memory test. The chapter covers the purpose of wafer matching during stacking and the difference between static and running repositories and their effect on stack yield and cost. Different wafer matching methodologies are explained, including matching algorithms and matching criteria. Fault models are developed for the faulty behavior of stacked memories using TSVs as bitlines and wordlines. Several 3D memory architectures with different test, repair, and redundancy architectures are introduced as comparisons for cost and yield benefits.

Chapter 3 provides a complete overview of the cutting-edge proposals for pre-bond testing of TSVs using built-in self-test (BIST). Electrical models are developed for TSV pillars and the defects that can arise during their manufacture. Several BIST architectures are compared to demonstrate their varying effectiveness in detecting TSV defects.

Chapter 4 provides an alternative solution for pre-bond TSV test through TSV probing. A discussion of present probe card technology and the limitations of TSV probing are presented, along with die and probe card architectures for enabling pre-bond parametric TSV test. The feasibility and accuracy of probing are analyzed in detail. An algorithm for reducing pre-bond TSV test time through testing multiple simultaneously TSVs through a single probe needle are also examined.

A variety of optimizations for reducing the overhead and test cost of pre-bond testing are introduced in subsequent chapters. Chapter 5 demonstrates how the architecture presented in Chap. 4 can be reused for pre-bond structural test. A complete analysis of the feasibility, speed, and drawbacks of the method is provided. Chapter 6 shows how novel applications of register retiming and new retiming flows can be used to minimize the delay overhead of the designs in Chaps. 5 and 4.

Chapter 7 discusses in detail the emerging test standards likely to govern the future of 3D SIC testing. Die-level wrapper will be introduced as a means of enabling a standardized pre-bond and post-bond test interface for dies to be used in a 3D stack. The JEDEC standard, which provides a high-speed bus interface for memory-on-logic stacks, will also be examined, including methods for memory test compatible with the standard.

Finally, Chap. 8 presents a unifying co-optimization for the 3D test-access-mechanism (TAM) and 2D TAMs to minimize test time through optimal test architecture design and test scheduling. The optimization can account for all possible post-bond test insertions. Furthermore, it covers a variety of 3D-specific test constraints, such as the number of dedicated test TSVs that can be added between any two dies.

This introduction will now continue with a general overview of circuit testing as it pertains to the new and exciting field of 3D SICs.

1.1 Basics of Testing

Testing of manufactured ICs by applying test stimuli and evaluating test responses is a necessary part of IC manufacturing flow. This ensures high wafer yields and that any given product functions to specification before being shipped to the consumer. Testing of ICs is an expansive topic and we will cover only the basics in this section. To begin, we will look at the four categories of testing: *verification testing*, *manufacturing testing*, *burn-in*, and *incoming inspection* [1]. We will then examine the differences between functional, structural, and parametric tests.

1.1.1 Categories of Testing

Verification testing is applied to a design prior to production. Its purpose is to ensure that the design functions correctly and meets specification. Functional and parametric tests, which will be discussed later, are generally used for characterization. Individual probing of the nets in the IC, scanning electron microscopes, and other methods not commonly performed during other tests, may also be used. During this time, design errors are corrected, specifications are updated given the characteristics of the design, and a production test program is developed.

Manufacturing, or production, test is performed on every chip produced [1]. It is less comprehensive than verification test, designed to ensure that specifications are met by each chip and failing those that do not meet standards. Since every chip must be tested, manufacturing tests aim to keep test costs low, which requires that test time per chip be as small as possible. Manufacturing tests are generally not exhaustive, aiming instead to have high coverage of modeled faults and those defects which are most likely to occur. In the rest of this book, we limit ourselves to manufacturing testing.

Even when manufacturing tests are passed and the chips function to specification, some devices will fail quickly under normal use due to aging and latent defects. Burn-in tests, which are often run at elevated voltages and temperatures, aim to push such devices to failure. This process removes chips that experience *infant mortality*.

Incoming inspection test takes place after devices are shipped and may not always be performed [1]. During incoming inspection, the purchaser of the device tests it once again before incorporating it into a larger design. The details of these tests vary greatly, from application-specific testing of a random sample of devices to tests more comprehensive than manufacturing tests, but the goal is to ensure that the devices function as expected prior to integration into a larger system when test becomes significantly more expensive.

1.1.2 Functional, Structural, and Parametric Testing

Functional tests aim to verify that a circuit meets its functional specifications [1]. Generally, functional tests are produced from the functional model of a circuit in which, when in a specific state and given specific inputs, a certain output is expected.

A benefit of functional testing is that the tests themselves can be easy to derive since they do not require knowledge of the low-level design. Since patterns used in verification testing are similar to functional tests, the patterns can be easily converted to functional patterns, which reduces costs in test development. Furthermore, functional tests can detect defects that are difficult to detect using other testing methods.

Despite these benefits, functional testing suffers from serious drawbacks [1]. In order to test every functional mode of a circuit, every possible combination of stimulus must be applied to the primary inputs. Thus, functional testing is prohibitively long unless, as is usually the case, a small subset of possible tests is used. This, however, leads to small defect coverage for functional tests. There is no known method for efficiently evaluating the effectiveness of functional test patterns.

While it has its drawbacks, functional test is usually included in product testing along with structural tests. Unlike functional testing, structural tests do not treat the circuit itself as a black box, instead generating patterns based on faults in specific areas of the netlist [1]. There are a number of fault models, the specifics of which are not discussed here, that can be used in generating structural tests. These models allow for the testing of specific critical paths, delay testing, bridging tests, and more. Producing structure-aware patterns leads to high fault coverage and generally reduces test time, especially compared to functional tests. Since specific models are used, structural tests can be more easily evaluated for fault coverage. The drawbacks of structural tests are that gate-level knowledge of the circuit is needed and that structural tests sometimes fail good circuits due to overtesting.

Parametric tests aim to test the characteristics of a device or a part thereof, and are generally technology-dependent [1]. Parametric tests can be separated into two categories—DC test and AC test. DC tests can include leakage tests, output drive current tests, threshold levels, static and dynamic power consumption tests, and the like. AC tests include setup and hold tests, rise and fall time measurements, and similar tests.

1.2 Design for Testability

Given the complexity of today's IC designs, comprehensive testing is impossible without specific hardware support for testing. *Design for testability* (DFT) refers to those practices that enable testing of VLSI ICs. In this section, we examine DFT techniques that enable the testing of digital logic circuits. Other methods are used for the testing of memory blocks, analog, and mixed-signal circuits, but these will not be discussed.

1.2.1 Scan Test

The vast majority of today's ICs are sequential circuits which rely on flip-flops and clocking to produce functionality. Testing of sequential circuits is very difficult because flip-flops must be initialized, must save states, and tend to have feedback. This greatly impacts both *controllability*, which is the ease with which a system can be placed in a desired state through its primary inputs, and *observability*, which is how easily internal states of the circuit can be propagated to primary outputs. In order to obtain controllability and observability at flip-flops, scan chains are commonly designed in the circuit.

Scan chains are based on the idea that an IC can be designed to have a specific test mode, separate from functional mode, to which it can be switched. While in test mode, groups of flip-flops are connected together to form a shift register called a *scan chain*. In order to do this, each flip-flop is designed as a scan-flop, which multiplexes between a functional input and a test input when in test mode. The test input is either a primary input for the circuit under test (CUT), or the previous scan-flop in the scan chain.

Scan chains enable controllability and observability at every scan-flop in the circuit. Test vectors are scanned into the scan chain from the inputs one bit at a time, with these bits shifted into scan-flops further in the chain. Test responses are then latched, and the response is scanned out through the output of the scan chain. Multiple scan chains can be tested in parallel to reduce test time, though each requires its own input and output in the CUT. There are overheads associated with scan test, including gate, area, and performance overheads, but the benefit to testability makes scan test common in most VLSI ICs.

1.2.2 Modular Test, Test Wrappers, and Test Access Mechanisms

A modular test approach to DFT separates a large system-on-chip (SOC), which may be composed of billions of transistors, into many smaller test modules. These modules are often partitioned based on functional groups, ranging from an entire core to analog circuit blocks. These modules may be considered stand-alone test entities from the rest of the modules in an SOC. This partitioning allows tests to be developed for each module independent of other modules in the SOC, which greatly reduces test generation complexity compared to developing tests for the top-level SOC. Furthermore, test reuse is possible if the same module is instantiated multiple times within the same SOC or between multiple IC designs. This also enables the purchase of modules from third-parties for incorporation into an SOC, such that tests for the module are provided and no knowledge of the implementation of the module is needed.

In order to easily test each module, the module must present a standardized test interface to the SOC integrator. It is for this reason that test standards, such as the IEEE 1500 standard [20], were developed. A *test wrapper* is used to provide controllability and observability at the boundary of the module. The test wrapper further enables appropriate test modes for the module and organizes scan chains in the module for test pattern delivery. We will briefly examine the 1500 standard as an example of wrapper design.

The 1500 test wrapper contains a wrapper instruction register (WIR) which can be configured to place the wrapper into specific functional or test modes. This is configured through the wrapper serial interface port (WSP) which contains the wrapper serial input (WSI), the wrapper serial output (WSO), wrapper clock (WRCK), wrapper reset (WRSTN), and other signals. A wrapper boundary register (WBR) is present into which patterns can be shifted for either external test of the logic between modules or internal test through the module's scan chains. A wrapper bypass register (WBY) allows test patterns or test responses to bypass the module on route to other modules in the SOC or to be output at the external SOC pins. Though all tests may be performed using the WSP, many wrappers also contain wrapper parallel in (WPI) and wrapper parallel out (WPO) busses. These consist of two or more bit lines for loading and unloading multiple internal scan chains at the same time. This design reduces test time, but requires more test resources.

In order to route test data from the external test pins to all modules in the SOC, a test access mechanism (TAM) is required. There are many ways to architect a TAM, including multiplexing, daisychain, and distribution designs [32]. Optimization of the TAM and test wrappers to minimize test time using limited test resources has been the subject of much research and will be discussed later in this book.

1.3 3D Integration Technology

A number of 3D integration technologies have been considered, but two main technologies have emerged—*monolithic* integration and *stacking* integration [4]. Although the research presented in the following chapters is based on a stacking approach, in which 2D circuits each with their own active device layer are bonded one on top of the other, we will first briefly examine monolithic technology.

Monolithic integration was proposed as an alternative to stacking, because the mask count and process complexity increases significantly with each stacked die. With monolithic 3D ICS, the processing for the creation of active devices is repeated on a single wafer, resulting in the 3D stacking of transistors. Since the devices and their wiring are processed on a single substrate, the added manufacturing complexities of thinning, alignment, and bonding and the need for through-silicon vias (TSVs) are nonexistent.

Because monolithic integration involves the creation of devices in the substrate above devices that have already been manufactured, significant changes in fabrication processes and technologies would have to take place [5]. The heat currently involved in active device processing is high enough to damage deposited transistors and melt existing wiring. Therefore, advances in low temperature processing technologies are needed. Some recent advances in these technologies [6, 7] have allowed monolithic designs to be realized in the laboratory [5].

Unlike monolithic integration, the active device layers in stacking-based integration are manufactured in separate substrates. Thus, each set of active device layer and associated metal layers are processed on wafers using current fabrication technologies, and substrates are then stacked one on top of the other to produce a 3D circuit. Because no significant changes are required in fabrication technologies, stacking-based integration is more practical than monolithic integration and has therefore been the focus of 3D research [4].

Stacking-based integration can be further separated into three categories based on the method of 3D stacking—wafer-to-wafer, die-to-wafer, and die-to-die stacking [4]. In wafer-to-wafer stacking, two or more wafers, each with many copies of a circuit, are stacked on top of one another and the resulting 3D stack is then diced to create the individual 3D stacked-ICs (SICs). In die-to-wafer stacking, two wafers are once again produced but one wafer is diced into individual dies, which are then stacked on the other wafer. More dies can be stacked after this process. In die-on-die stacking, wafers are diced into individual dies and then stacked. Die-to-die stacking is desirable as it allows for testing of individual die prior to being introduced to a stack. Aside from being useful for increasing stack yield by discarding bad dies, die-to-die stacking allows for binning of dies to match a stack for performance and power.

Some method must exist in a 3D SIC for dies in the stack to interconnect to one another. A number of methods have been proposed for this interconnection, including wire bonding, microbump, contactless, and TSVs [25]. Wire bonding makes connections between the board and stack or between dies themselves, though wires can only be on the periphery of the stack. Wire bonding thus suffers from low density, a limit on the number of connections that can be made, and the need for bonding pads across all metal layers due to the mechanical stresses of the external wires. Microbumps are small balls of solder or other metals on the surface of the die that are used to connect dies together. They have both higher density and lower mechanical stress than wire bonding. Microbumps do not, however, reduce parasitic capacitances because of the need to route signals to the periphery of the stack to reach destinations within it. Contactless approaches include both capacitive and inductive coupling methods. Though resulting in fewer processing steps, manufacturing difficulties and insufficient densities limit these methods. TSVs hold the most promise as they have the greatest interconnect density, though they also require more manufacturing steps [28]. We assume stacking-based integration, whether wafer-to-wafer, die-to-wafer, or die-to-die, with TSVs for all chapters of this book.

TSVs are vertical metal interconnects that are processed into a substrate at some point during manufacture. In a front-end-of-the-line (FEOL) approach, TSVs are implanted into the substrate first, followed by active devices and metal layers. In a back-end-of-the-line (BEOL) approach, the active devices are processed first, followed by the TSVs and metal layers [3]. In certain limited stacking approaches, TSVs may also be manufactured later in the process (post-BEOL). This can be done before bonding (vias first) or after bonding (vias last).

In all TSV manufacturing approaches, the TSVs are embedded in the substrate and need to be exposed. TSVs are exposed through a process called "thinning", in which the substrate is ground away until the TSVs are exposed. This step results in dies that are much thinner than conventional 2D substrates, and are thus quite fragile and are commonly attached to carrier wafers before 3D integration. In order to be attached to other dies in a 3D stack, a die must go through "alignment" and "bonding". During alignment, the dies are carefully placed such that their TSVs make direct connections to one another. During bonding, the dies are permanently (current technology does not support "unbonding" of dies) connected to one another, making contact between TSVs. Bonding can be done through a variety of methods, including direct metal-to-metal bonding, direct bonding of silicon oxide, or bonding with adhesives [3]. The processes of alignment and bonding continue until all thinned dies are integrated into the 3D SIC.

There are two different approaches to stacking dies—face-to-face and face-to-back. In face-to-back bonding, the outermost metal layer of one die (the face) is connected to the TSVs on the substrate side (the back) of the other die. Face-to-back allows for many die layers to be bonded in a stack. In face-to-face stacking, the faces of two die are connected to one another. This can reduce the number of TSVs needed for connections, but can support only two dies in a stack unless face-to-back bonding is used for other dies. Though back-to-back bonding is conceivable, this is not a common approach.

To better illustrate 3D integration, Fig. 1.1 shows an example of a 3D SIC. This is an example of a stack with two dies bonded face-to-face. Only Die 2 has TSVs to connect to external I/O bumps, and so it is thinned while Die 1 is not. A heat sink is attached to the outside of the stack.

Commercial products using 3D stacks with TSVs are available [8, 9], but are limited to stacked memories. Due to the relative ease of test and repair for memory devices, built-in self-repair and wafer matching techniques can result in significantly high product yields. Faulty stacks that cannot be repaired are discarded. In order to realize the full potential of 3D integration—memories stacked on cores, 3D cores, mixed technology, and the like—testing techniques have to be developed for use during 3D manufacturing.

1.3.1 3D Testing

Compared to the testing of 2D ICs, 3D SICs introduce many new challenges for testing. Yield loss for each die in a 3D SIC is compounded during stacking, so stacking of untested die leads to prohibitively low product yields. This motivates the need for *pre-bond* testing, or the testing of dies prior to being bonded to a 3D stack. This allows for the stacking of die that are known to be defect-free and also enables

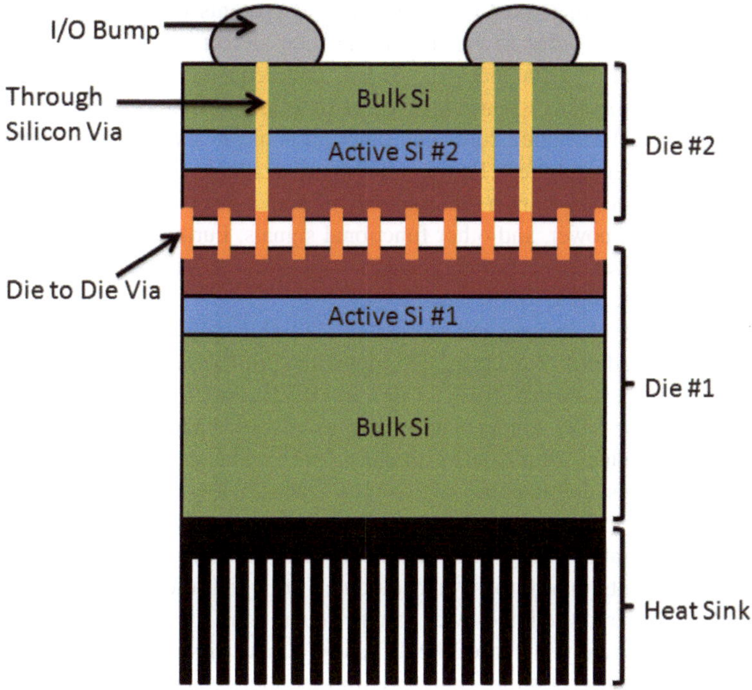

Fig. 1.1 Example of a face-to-face bonded 3D SIC with two dies

die-matching, so that dies in the same stack can be chosen based on metrics such as speed or power consumption. It is also important to perform *post-bond* tests—testing of either a partial stack to which all dies have yet to be bonded or testing of the complete stack. Post-bond testing ensures that the stack is functioning as intended and that no errors have been made or new defects have been introduced during thinning, alignment, and bonding.

Pre-bond testing of dies offers many design-for-test (DFT) challenges. First, thinned wafers are far more fragile than their non-thinned counterparts, so a small number of contacts must be made during probing and low contact-force probes are a necessity. Because of design partitioning in 3D stacks, a die may contain only partial logic and not completely functional circuits. Currently, this limits the number of tests that can be applied to a circuit with partial logic, though future breakthroughs may make these dies more testable. TSVs also present problems of pre-bond testing because high densities and small sizes make them difficult to probe individually with current technology. Limitations on dedicated test TSVs, oversized test pads for probing and test signals, and the availability of I/O pins only through one end of the stack make design and optimization tools important for proper test-resource allocation. Such resource constraints are present post-bond as well.

Like pre-bond testing, post-bond testing also presents difficulties not present in 2D IC testing. In order to ensure that no new defects are introduced during stacking, testing of a partial stack is needed. This requires a test architecture and appropriate optimizations to ensure that test time remains small and partial stack tests are possible. Embedded cores and other parts of the stack may span multiple dies, further complicating test. Few test TSVs are available to the stack, since each additional TSV needed restricts the number of active devices and most TSVs are needed for clocks, power, and other functional signals. Furthermore, limitations in test access are present from few dedicated test pins that provide test access through only one end of the stack.

In [35], a die-level wrapper and associated 3D architecture are presented to allow for all pre-bond and post-bond tests. This approach proposes die-level wrappers and it leverages current standards, IEEE 1149.1 and IEEE 1500. In addition to functional and test modes, die-level wrappers allow bypass of test data to and from higher dies in the stack and reduced test bandwidth during pre-bond tests. While this provides a practical look at test architectures in 3D-SICs—what 3D wrappers and TAMS may be like—it offers no insight into optimization and test scheduling.

Other considerations may also impact the choices made for 3D SIC testing. Thermal constraints may limit which modules on which dies may be testable at any given time, or require low-power pattern generation. The stacking of substrates and oxide layers greatly increases the difficulty of heat dissipation, particularly for die farther from the heat sink. This leads to complications especially during test, when the stack is often heating faster than under functional operation.

TSVs themselves may also impact the operation of surrounding devices. Since a "keep-out" area around each TSV may have a significant impact on the area overhead of 3D integration, active devices may be placed close to TSVs during layout. The stresses in the semiconductor crystal caused by the presence of a TSV may alter electron and hole mobility near the TSV. Depending on the distance and orientation of transistors near TSVs, this may cause them to run slower or faster than normal. Such considerations are important for test development.

1.3.2 Conclusion

This chapter has provided an overview of the testing challenges that must be overcome before 3D SICs can be widely adopted by industry. Optimization techniques are needed to make the best use of limited resources, both in terms of test access and test scheduling. New methods and breakthroughs are needed to make TSV testing economical and to enable the testing of partial logic. Each chapter from this point on will provide in-depth examination of individual 3D SIC testing topics as well as describe prior work that allows understanding of research advances in an appropriate context.

Chapter 2
Wafer Stacking and 3D Memory Test

2.1 Introduction

In any 3D manufacturing and test flow, it is necessary to determine which tests should be performed and when, as well as how stacking will take place in order to minimize cost. This determination is necessary to ensure suitably high compound stack yields, or the yield for stacking subsequent tiers on a stack. This chapter will examine two related issues—the stacking process, in particular the benefits and cost of wafer sorting, and architectures for 3D memory test and yield assurance in 3D stacked ICs. Pre-bond test of logic, memory, and interconnects is necessary to enable wafer matching, and memory test is crucial especially for memory-on-memory stacks where wafer stacking is more likely to be used due to higher yields through repair solutions.

The rest of this chapter is organized as follows. This section introduces and motivates the concepts of wafer matching and 3D memory test and repair. Section 2.2 provides a mathematical model for estimating yield and cost of utilizing static repositories for wafer matching and demonstrates the yield improvements that wafer matching offers for stacked ICs. Section 2.3 expands the discussion of Sect. 2.2 to running repositories and evaluates the benefits of different matching flows. Section 2.4 discusses fault modeling for stacked memories with resistive-open defects on TSVs that are utilized as bitlines and wordlines between memory dies. Finally, Sect. 2.5 examines three different stacked memories with varying test, repair, and redundancy strategies and their impact on stack yield and cost. Section 2.6 concludes this chapter.

2.1.1 Wafer Stacking Methodologies

The stacking of dies to create a 3D SIC can generally occur in one of three ways—wafer-to-wafer (W2W) stacking, die-to-wafer (D2W) stacking, or die-to-die (D2D)

B. Noia and K. Chakrabarty, *Design-for-Test and Test Optimization Techniques for TSV-based 3D Stacked ICs*, DOI 10.1007/978-3-319-02378-6_2,

stacking. In W2W stacking, two wafers are bonded to one another, with dies in the same position on each wafer forming the stack. W2W bonding benefits from the highest manufacturing throughput of all of the stacking methods due to the need for only a single alignment step to bond two tiers of many stacks at the same time. Furthermore, it is easier for machines to handle and align large wafers than to do the same for dies when the dies are relatively small. For processes with particularly high wafer yields, pre-bond test may not need to be performed and unsorted wafers can be stacked to keep test costs low. The drawback of W2W bonding is that there is little flexibility as to which dies are bonded together. Due to this, W2W stacking is expected to have the worse compound stack yield out of the three methods as it may not be possible to prevent a bad die and a good die being bonded together. In addition, dies cannot be easily matched to one another for power, speed, thermal, or other design considerations for stacking. W2W bonding requires that dies on each wafer are the same or similar size.

In D2D bonding, wafers are diced and dies are tested prior to bonding. Individual dies can then be binned based on design considerations to best match one another, and faulty dies are discarded altogether. In this way, only matched, known-good-dies are bonded to one another. This can lead to very high compound stack yields. The drawback to D2D stacking is that manufacturing throughput is low, as more alignment steps are required compared to W2W bonding. Furthermore, it may be difficult to achieve high alignment accuracy in a timely manner when particularly small dies are handled.

D2W bonding offers a middle ground between W2W and D2D bonding that can offer similar yields to D2D bonding. In a D2W process, pre-bond test is performed on wafers that will make up two tiers of the stack. The wafers for one of the tiers is diced, then dies are matched and bonded to dies on the wafer for the second tier. It is important to note that a test and manufacturing process for a complete 3D stack may make use of multiple bonding methods. For example, the first two dies of a 3-die stack may be bonded using a D2W process, and then the third die is bonded to the stack in a D2D bond.

The rest of this chapter pertaining to stacking methodologies will focus on W2W bonding and methods to increase yield through wafer sorting and wafer repositories. In order to improve W2W stack yields beyond random wafer stacking, pre-bond test can be performed to create a *wafer map* for the dies on a wafer. The wafer map contains the location of each die and whether it passed or failed its pre-bond testing. In order to benefit from the wafer map, repositories are used to store more than one wafer for each tier of a 3D stack. A wafer matching algorithm is then utilized to match wafers in one repository to their best match in one or more other repositories to maximize the compound stack yield after bonding. Though a number of considerations can be taken into account during wafer sort, generally wafers are chosen to maximize the number of good dies and bad dies on one wafer that will be bonded to a good die or a bad die, respectively, on a second wafer. The 3D wafer matching problem of matching good and/or bad dies between wafers has been shown to be NP-hard [65], but this chapter will not focus on wafer sorting algorithms and instead focus on W2W stacking methodologies themselves.

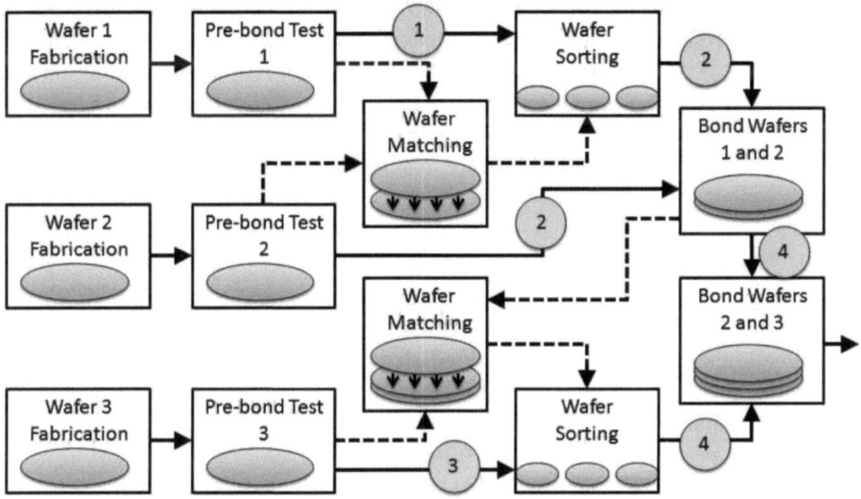

Fig. 2.1 An example of a W2W stacking flow for a 3-tier 3D SIC

2.1.2 W2W Stacking and Wafer Matching

Figure 2.1 shows an example W2W stacking flow for a 3-die stack. First, wafers are fabricated separately for each die in the stack, with Wafer 1 containing the dies for the lowest stack tier, Wafer 2 containing dies for the second stack tier, and so forth. These wafers then undergo pre-bond test to create the wafer map of good and bad dies. Wafers 1, 2 and 3 are then placed in separate repositories, and wafer fabrication and test continues until the wafer repositories all contain an appropriate number of wafers. Once the repositories are filled, wafer matching and sorting can take place.

To begin bonding, first the wafer maps for the wafers in the Wafer 1 repository and the Wafer 2 repository are sent to a computer to be matched via a wafer matching algorithm. The resulting matching data is sent to a wafer sorter along with the Wafer 1 repository at Point 1 in Fig. 2.1. The Wafer 1 repository is sorted to align the wafers with their matched wafers in the Wafer 2 repository. At Point 2, both the Wafer 1 and Wafer 2 repositories are sent to be bonded together to form a partial stack.

At this point, it is possible to perform a partial stack test to update the wafer map of Wafer 2 to account for faulty stacks that may arise by bonding or mechanical failures. In the flow of Fig. 2.1, however, no partial stack tests takes place, and instead the wafer maps for Wafer 1 and Wafer 2 are combined such that a new wafer map contains the bad dies from both Wafer 1 and Wafer 2. This new wafer map is sent to be matched with the wafer maps in the Wafer 3 repository. Then, at Point 3, the Wafer 3 repository is sent to be sorted based on the match results. Finally, at Point 4, W2W bonding is performed to bond Wafer 3 to the stack of Wafer 2 and Wafer 1. From here, a stack test can be performed before or after dicing the stacks apart before the stacks go on to assembly, packaging, and final test.

Figure 2.1 provides a general overview of a W2W process, but there are a number of steps in the process that can change the overall cost and yield of W2W stacking [66]. First, the repository can either be *static* or *running* depending on the repository capability. Second, the *wafer matching process* determines the level of granularity taken into account when matching wafers. Finally, the *matching criteria* determines by what standards wafers are matched to one another.

2.1.2.1 Static and Running Repositories

It would be exceedingly costly to match all the wafers produced in a manufacturing run with all other wafers. By using limited-size repositories, only a small number of wafers must be matched to one another, thereby reducing the size of the matching problem to a manageable level. Two types of repositories—static or running—can be used for wafer matching.

Consider a repository that can hold m wafers out of a production run of e ($m \leq e$) wafers. In a *static* repository, the repository begins by being filled with m wafers. The wafers in this repository are matched with one or more other repositories, and all repositories are then emptied of the matched wafers. Only after being emptied completely is the repository replenished, and this procedure is repeated e/m times.

In a *running* repository, the repository begins by being filled and wafer matching takes place similarly to the static repository. However, after each matching step, only the best-matched wafer in the repository is removed. After the wafer is removed, it is immediately replaced by a new wafer and matching is performed again with a full repository. This process occurs e times.

Static repositories offer less choice among wafers during wafer matching when compared to running repositories. During the first wafer matching step in a static repository, there are m wafers to choose from. For the second match, there are $m - 1$ wafers to choose from, and so forth. With each match, there is less freedom among the other choices. A running repository, by comparison, generally maintains a full repository, so there is always m wafers to choose from. In practice, however, *repository pollution* can occur in running repositories that also decreases choice in wafer matching. Repository pollution refers to the fact that particularly poor wafers may not be matched to another wafer over many matching iterations, effectively causing them to occupy space in the repository and reducing the number of good wafers that can be chosen from.

Compared to running repositories, static repositories are generally easier to implement in a real manufacturing process. In an actual production line, wafer containers are utilized to move groups of wafers from one machine to another. These can be treated as static repositories in wafer matching, sorting, and stacking. In order to create a running repository, the wafer container must either move many times from the bonding machine back to the wafer production line so that wafers can be added to it, or extra containers must be transported with the repository container in order to provide a source for additional wafers.

2.1.2.2 Wafer Matching Processes

The wafer matching process that can be used is determined by the number of repositories that are used simultaneously during matching and the number of wafers within each repository that are considered during matching. There are two matching dimensions that depend on the number of repositories used:

- *Layer-by-layer* (LbL) refers to the process shown in Fig. 2.1. In LbL matching, only two repositories are considered at any given time. For a stack of three dies, this means that repositories containing the wafers of the first two stack tiers are matched against one another and the wafers then bonded. To complete the stack, the repository of the partial stack is matched against the repository of the wafers containing Die 3, and then this wafer is bonded to the partial stack. This matching process is iterative for each bond in the stack.
- *All-layers* (AL) refers to a matching process that considers all the repositories for each tier in the stack at the same time. This process is complete and may be capable of better results than an LbL process, but is also computationally more difficult.

There are also two wafer matching dimensions depending on the number of wafers considered in each repository simultaneously:

- *Wafer-by-wafer* (WbW) is a greedy matching process whereby only the best-matched wafers between the repositories are considered at a time. These best-match wafers are then removed from the repository for purposes of matching, and the best-match is chosen from the remaining wafers. This process continues until the repositories are empty.
- *All-wafers* (AW) is an exhaustive matching process whereby every wafer in every repository visible to the process is considered simultaneously. In other words, all possible matching outcomes are checked for each of m wafers across two or more repositories. Wafers are then matched to each other based on some criteria, for example maximizing the total expected compound stack yield across all wafers.

The above dimensions of the wafer matching process can be combined to form five possible complete processes. Figure 2.2 shows the five processes and how they arise from the matching dimensions. The possible matching processes are explained below:

- LbL→WbW is iterative over two repositories at a time. In each iteration, the two best-matched wafers in each repository are chosen and removed from consideration. Each successive step matches the remaining wafers, which after the first step is either $m - 1$ for a static repository or m again for a running repository. This continues until all wafers are matched, and then the process moves on to the repository for the next stack tier and so on.
- WbW→LbL is similar to LbL→WbW except that after choosing the first best-match wafer pair from the first two repositories, it then attempts to match the pair to the best match from the next stack tier's repository and so forth until one

Fig. 2.2 The possible wafer matching processes across two dimensions

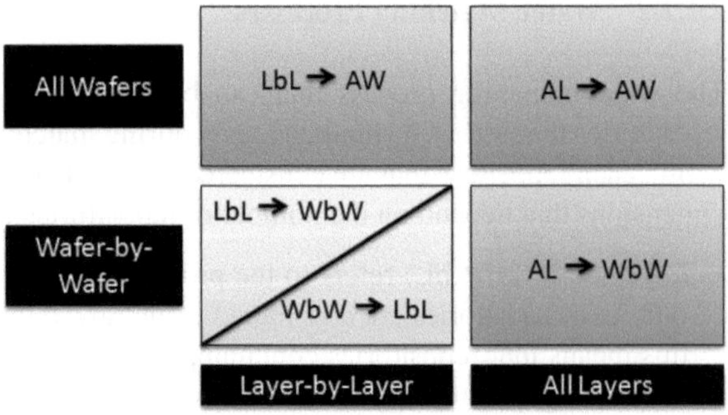

wafer is removed from all tier repositories. This continues for one wafer from each repository until all repositories are empty.

- LbL→AW is also similar to LbL→WbW in that it considers only two repositories per iteration. However, for each iteration, all the wafers in one repository are matched to all the wafers in the other. This is a process that utilizes static repositories, as the repository is completely matched and emptied before replenishing its wafers. This matching process continues to the next repository, completely matching all wafers, and so forth until the stack is complete.
- AL→WbW considers all repositories simultaneously for all n tiers in a 3D stack. The best-matched n wafers are selected such that one wafer is removed from each repository. The process is repeated with the next best matching wafers removed from each repository and so on until the repositories are empty. This process is compatible with both static and running repositories.
- AL→AW is the most exhaustive of the possible matching processes. In this case, all m wafers across all n repositories are matched simultaneously depending on the matching criteria, and all repositories are emptied simultaneously before being replenished. This process can only be used with static repositories and requires significant computation.

2.1.2.3 Wafer Matching Criteria

There are a number of different matching criteria that can be chosen for matching based on the wafer maps for each repository. The following three criteria are examples of useful metrics, but they are by no means exhaustive:

- *Max(MG)* is a criterion where wafers are matched to one another based on maximizing the number of good dies that align with one another between the wafers.
- *Max(MF)* is a criterion where wafers are matched based on maximizing the number of faulty dies that align between the wafers.
- *Min(UF)* is a criterion where wafers are matched in order to minimize the number of unmatched faulty dies between the wafers. In effect, this improves stack yield by stacking wafers such that the number of good dies that are bonded to bad dies is minimized.

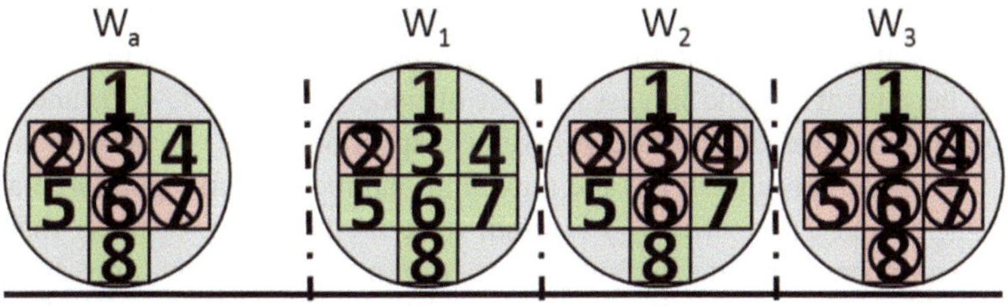

	W_1	W_2	W_3
Matched Good Dies	4	3	1
Matched Faulty Dies	1	3	4
Unmatched Faulty Dies	3	1	3

Fig. 2.3 An example of wafer matching for three different criteria

Note that combinations of these matching criteria can be used if weights are applied to each criterion depending on importance. These criteria can be mathematically modeled if we define a vector F for the fault map of a wafer in the repositories to be considered, where F is a string of binary values that denote whether a die in each location on a wafer is good or bad. A function $G(F_i)$ can be defined to denote the number of bad dies on a wafer with fault map F_i. Then we can model the criteria as follows (note that $0 \leq i, j \leq m$, where m is the repository size, and & refers to a bitwise logical AND):

$$Max(MG) = \max_{i,j}\{G(\bar{F}_i \& \bar{F}_j)\} \qquad \forall i, j \tag{2.1}$$

$$Max(MF) = \max_{i,j}\{G(F_i \& F_j)\} \qquad \forall i, j \tag{2.2}$$

$$Min(UF) = \min_{i,j}\{G(F_i \oplus F_j)\} \qquad \forall i, j \tag{2.3}$$

In an AW matching process using static repositories, it does not matter which criterion is utilized, since the exhaustive nature of the process will provide the same compound yield regardless. However, the chosen criterion will impact compound yield if a WbW matching process is used or if running repositories are used.

Figure 2.3 demonstrates the differences between the wafer matching criterion on a static repository. One wafer, W_a is being matched against three possible wafers W_1 through W_3. There are eight dies on each wafer numbered 1 through 8 such that die 1 on wafer W_a will be matched with die 1 on one of the other wafers and so on. Each wafer has a certain number of faulty dies—in the case of W_a, dies 2, 3, 6, and 7 are denoted as being faulty. The table at the bottom of the figure labels how many dies match given certain criteria. For example, there are four good dies that match between W_A and W_1, three between w_a and w_2, and one between W_A and W_3.

Using the example of Fig. 2.3, it is clear that each matching criterion results in a different wafer match. For the *Max(MG)* criterion, wafer W_1 will be matched with W_a because it has the largest number of good dies (dies 1, 4, 5, and 8) that correspond with good dies on the W_a wafer map. If *Max(MF)* is used, then wafer W_3 will be chosen because it has the maximum number of corresponding bad dies (dies 2, 3, 6, and 7). If *Min(UF)* is used, then W_2 is chosen because there is a minimum number of unmatched faulty dies between the two wafers (only die 4). The corresponding compound stack yields of using Max(MG), Max(MF), or Min(UF) in this example without considering stacking defects will be 50%, 12.5%, and 37.5%. Despite Max(MG) having the highest yield in this particular matching case, the other criteria may result in higher yields overall when future wafer matches are considered.

2.1.3 3D Memory Architectures and Memory Test

In most implementations of 3D stacked DRAM memories, whether they be memory-on-logic or memory-on-memory stacks, memory arrays are placed on stack layers different from peripheral logic such as decoders and sense amplifiers [67]. This is because DRAM cells are manufactured using an NMOS process, whereas peripheral logic and computing cores are fabricated in a CMOS process. Separating the technologies between dies capitalizes on the benefits that 3D stacks offer for heterogeneous integration, such as reduced manufacturing cost and complexity. Including memory arrays on separate dies from logic also allows for better layer optimizations, for example optimizing the memory array for density and logic for speed.

It is possible to reuse a 2D-style memory layout in a 3D stack, for example by including peripheral logic on every DRAM layer and utilizing TSVs as buses to link the memory layers to a logic layer. While such a design would reduce the number of TSVs needed while still reducing memory access time, it does not benefit from heterogeneous integration and does not increase bandwidth between memory structures themselves. By comparison, Fig. 2.4 gives an example of a memory-on-logic stack that takes advantage of the benefits of 3D integration. The figure shows one logic die on the lowest level of the stack with two memory dies bonded to it.

In Fig. 2.4, the peripheral logic, memory controllers, and logic cores all belong to the bottom logic die while the other two dies contain DRAM memory arrays. TSVs act as bitlines and wordlines to connect the array cells to the periphery logic on the bottom die. Decoders and multiplexers can be added to the memory array dies to share a single TSV between multiple wordlines and to select one of several bitlines to connect to a peripheral sense amplifier. Similar designs have been proposed in the literature [68, 69], and Tezzaron Semiconductor has manufactured such designs in silicon [70].

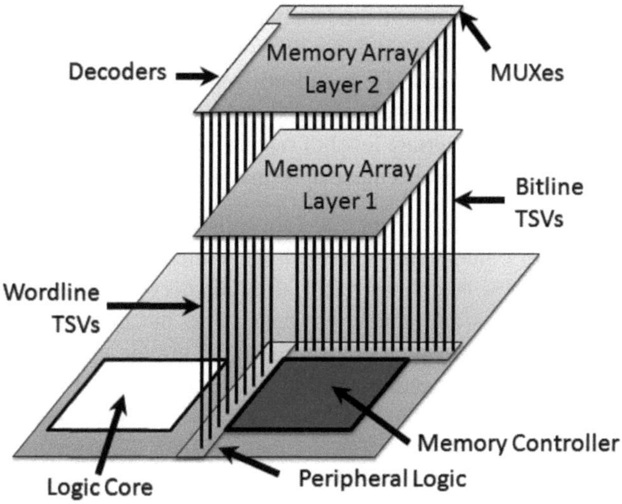

Fig. 2.4 An example of a memory-on-logic stack with two tiers of DRAM arrays

The design of Fig. 2.4 is an example of a 3D memory architecture referred to as *cell arrays stacked on logic* [71]. It is one of three overarching schemes for designing 3D memories. These three memory architectures are defined as follows:

- *Stacked banks* refers to a memory architecture whereby memory is partitioned bank by bank, such that entire banks are stacked on top one another. Each bank contains all the peripheral logic, the memory arrays, and other circuits (address decoder, write drivers, etc.) that make up a complete memory system. This 3D memory architecture is at the coarsest granularity, but leads to reduction in wire lengths that reduces power and memory access delay when reading or writing memory.
- *Cell arrays stacked on logic* is the type of memory architecture presented in Fig. 2.4 and is the predominant design for 3D memories. The peripheral logic is included on a logic-only die, and two or more memory array dies are added to the stack. As mentioned earlier, this method allows for separate optimization of the logic and memory array dies, and this architecture can be used for both DRAM and SRAM. Power and latency reductions are realized by utilizing TSVs as wordlines and bitlines. This memory architecture can split memory arrays using two methods. In a *divided-columns* design, bitlines are mapped onto separate dies. In a *divided-rows* design, the wordlines are separated between dies. Unless decoders are used, one wordline maps to one TSV.
- *Intra-cell (bit) partitioning* refers to a 3D memory architecture where memory cells themselves are split across multiple dies. This architecture suffers from the difficulty of integrating relatively small memory cells with TSVs, but may be feasible for certain designs [72].

Testing of 3D memories is not fundamentally different from testing 2D memories. Many concepts, designs, and applications will be familiar. For example, memory built-in self-test (MBIST) is still utilized to locate faulty memory cells, and built-in self-repair (BISR) mechanisms along with built-in redundancy-analysis (BIRA) circuits allow for a certain level of repairability. There are two predominant BISR architectures for 2D designs that can be mapped to 3D integrated memories:

- *Decoder redirection BISR* is an architecture consisting of four parts. One or more redundant rows or columns are included in the memory blocks to be used as a replacement for a row or column with faulty memory cells. A relatively simple MBIST circuit creates and sends test patterns to the memory cells. A BIRA circuit analyzes the test responses form the BIST circuit and determines how to utilize the redundant rows and/or columns for repair. A fuse macro permanently stores the routing information from the BIRA and updates a decoder to utilize the redundant memory in place of faulty memory.
- *Fault-cache BISR* also utilizes MBIST and a BIRA circuit, but includes a fault cache instead of a decoder and uses global redundant units (GRUs) in place of redundant rows and/or columns. The GRUs are separate from the individual memory blocks and so can be used to repair any memory block. The BIRA results are stored in the fault cache. Data that would be stored in a faulty memory cell is instead stored in a GRU, and the fault cache mediates data access stored in the GRUs.

Despite the similarity between 2D and 3D memory test and repair, crucial differences exist. The impact that resistive-open faults have on TSV-based wordlines and bitlines in 3D memories differ from the same defects in 2D memories using traditional interconnects. This is because capacitive coupling between high-density TSVs is more significant than between 2D interconnects, so aggressor-victim interactions between TSVs are more severe. Detecting open faults is made more difficult in 3D circuits because the faulty behavior of a TSV depends on voltage-dependent parasitics from neighboring TSVs, on whether a faulty TSV is subject to a read or write operation, and on the effects of coupling capacitance with its surroundings. Furthermore, traditional address fault (AF) models for 2D circuits do not map directly to open defects in the address decoders of 3D memories.

Differences between 2D and 3D memories with respect to repair are due to the types of 3D memory designs and the additional freedom in types of redundancy and placement of test circuits provided by 3D integration. Generally, there are three types of redundancy that are possible in a 3D memory architecture [71]:

- *Intra-layer redundancy* is most analogous to 2D redundancy architectures. In such a design, each memory die in the stack contains its own redundant resources that can be used to repair faulty memory cells on the die. Each memory die in the stack does not share its redundant resources with other dies in the stack.
- *Inter-layer redundancy* refers to a design where each die may or may not contain its own redundant resources. If a die does contain its own resources, then it shares these resources with other dies and has access to the resources on other dies.

In this way, a die can utilize redundant resources on another die if it cannot repair all of its faulty memory cells with its own resources. If each die does not have its own redundant resources, then it accesses and shares redundant resources on another die in the stack.

- *Layer redundancy* refers to a repair architecture where redundant resources exist at the die or wafer level. In other words, any given 3D memory stack may have one or more dies devoted to redundant memory arrays that can be utilized for repair. If a memory die is unable to repair itself due to a large number of faulty memory cells, then the entire die must be replaced by a redundant memory die.

In this chapter, we examine the new fault models required to accurately test 3D memories in Sect. 2.4. These fault models will be derived for the cell arrays stacked on logic model of 3D memories. In Sect. 2.5, three different redundancy designs will be examined for 3D memory stacks. Two designs utilize inter-layer redundancy, either with dies sharing resources between each other or all resources placed on the predominantly logic die with the BISR circuit. The third design will examine layer redundancy with spare dies.

2.2 Test Cost and Yield Benefits of Matching Static Repositories

This section evaluates the yield benefits and impact on test cost of stacking tested wafers from matched static repositories [63]. This is a W2W stacking method that occurs after pre-bond test and the creation of a wafer map describing which dies are good and bad on each wafer. The wafer maps in each repository are compared against one another, and the wafers in one repository are sorted to best match good dies on the wafers in the other repository before stacking. This process is repeated, considering the top wafer of the stack matched against the repository containing the wafer of the next die in the stack, until the entire stack is assembled.

There are four factors that impact stack yield and the cost of test that will be considered in this section. The first factor is the yield of a wafer. The yield of a wafer is given as a percentage and refers to the number of dies on that wafer that pass known-good-die (KGD) test. In other words, a 95% yield means that 95 % of the dies on the wafer are expected to be good dies. The distribution of the good dies and the bad dies across the wafer is important for wafer matching.

The second factor that will be considered is the number of dies in a stack. As the size of the stack increases, the *compound yield* of the stack is expected to decrease, especially in a W2W bonding scenario. Compound yield refers to the likelihood of a stack being considered good as each subsequent die is added to it. For example, if all dies in a stack are identical and each wafer has a yield of 95%, then a two-die stack would have a compound yield of 90.3% assuming that the stacking process has a 100% yield. A three-die stack would have a compound yield of 85.7%, and so on. These yields would be worse if each stacking process has a chance of causing a failure of the stack, and this would also have a greater impact on compound yield the more dies were in a stack.

The third factor is the number of dies per wafer. The number of dies per wafer is a function of both the area of each die and the size of the wafer. Smaller dies can reduce manufacturing cost by utilizing more of the wafer area, while larger dies lead to more unused space around the edges of the wafer. Furthermore, for a given defect density, smaller dies have a significantly higher yield.

The last factor that will be considered is the size of the wafer repositories. It is assumed that the repositories for the wafers of each die in the stack are the same size. Larger repositories are expected to improve the quality of matches between wafers, but can lead to greater test costs due to the requirement of more wafer storage and larger, more complex sorting machines.

2.2.1 *Calculating Yield for Static Repository Matching*

Utilizing the factors discussed at the beginning of this section, a recursive model is created for the expected compound yield Y of the stack [63]. Yield loss from defects caused by stacking will be factored in later, and for the time being is ignored. Furthermore, the effect of the repository size m will be disregarded for the time being, such that $m = 1$. Larger values for m will be factored in later.

The variables that will be used in the model are first defined. Each wafer contains d dies, and each stack is made up of n dies. For each bonding step in wafer stacking, we examine only two wafers regardless of the stack tier—a wafer a and wafer b. There are g_a faulty dies on wafer a and g_b faulty dies on wafer b and the wafer maps are known, because KGD test has already been performed. In the yield model, which wafer is referred to as a and which wafer is referred to as b is interchangeable, so long as $0 \leq g_a \leq g_b \leq d$. If it happens that g_a is greater than g_b, then wafers a and b are switched for the purpose of the model. One wafer refers to the wafer being added to the stack that is part of the sorted set of wafers, and the other wafer refers either to the wafer containing the dies lowest in each stack if no stacking has occurred yet, or the highest wafer currently bonded to the stack in the case that a partial stack exists.

2.2.1.1 Yield for Repository Size $m = 1$

The function $y(i)$ is equivalent to the compound yield of stacking wafer a and wafer b, where i $(0 \leq i \leq g_a)$ faulty dies are bonded together as part of the same stack. This function can be defined for each stack step and is calculated as:

$$y(i) = \frac{\max(d - g_a - g_b + i, 0)}{d} \tag{2.4}$$

The max function subtracts the number of bad dies in wafers a and b from the total number of dies d on a wafer. This would be the number of stacks with two good dies bonded together were there no overlap of faulty dies. This overlap i is added back in to prevent double-counting of faulty stacks. If for some reason the result of the algebra is negative, the max function and the yield become 0. The max function is divided by d to obtain the ratio of faulty stacks to total stacks for the wafer stacking step. More complete definitions of g_a and g_b will be described later to account for faulty stacks created in previous stacking steps.

We define another function $p(i)$ that corresponds to the probability that, for each possible value of i, exactly i bad dies will be bonded together during a bonding step. The function $p(i)$ is defined as:

$$p(i) = \binom{g_b}{i} \cdot \binom{d - g_b}{g_a - i} / \binom{d}{g_a} \tag{2.5}$$

The first choose term is the number of possible matches of i bad dies in wafer a to g_b bad dies on wafer b. This is multiplied by a second choose term that is equivalent to the number of possible matches between the remaining $g_a - i$ bad dies on wafer a and the good dies $d - g_b$ on wafer b. To obtain the appropriate probability, this product is taken as a ratio over the number of possible matches of all bad dies on wafer a with all dies on wafer b. It is always the case that:

$$\sum_{i=0}^{g_a} p(i) - 1 \tag{2.6}$$

The expected compound yield of the stack can now be defined as

$$Y = \sum_{i=0}^{g_a} (y(i) \cdot p(i)) \tag{2.7}$$

To demonstrate how to find Y, we use an example of two wafers as shown in Fig. 2.5 with $d = 6$, $f - a = 2$, and $f_b = 3$. The repository size is $m = 1$, so these two wafers must be stacked. We calculate $y(i)$ for the three possibilities of no bad dies ($i = 0$), one bad die ($i = 0$), or two bad dies overlapping ($i = 0$):

$$y(0) = \frac{1}{6}, \quad y(1) = \frac{1}{3}, \quad y(2) = \frac{1}{2}$$

We then calculate $p(i)$ for our range of i values, beginning with the probability of exactly zero bad dies being bonded together,

$$p(0) = \binom{3}{0} \cdot \binom{6-3}{2-0} / \binom{6}{2} = \frac{1}{5},$$

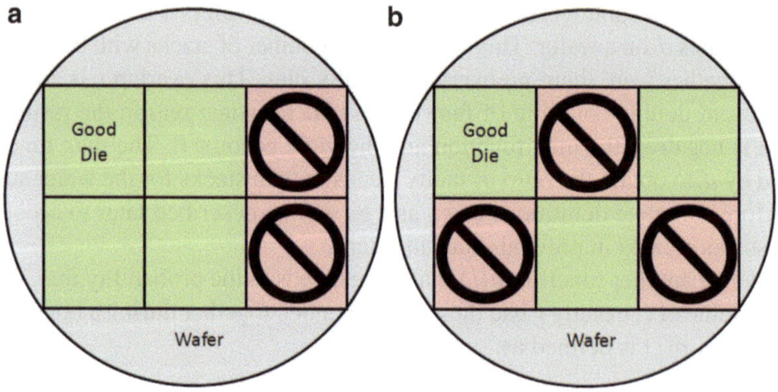

Fig. 2.5 Example wafers a and b when $d = 6$, $g_a = 2$, and $g_b = 3$

then for $i = 1$,

$$p(1) = \binom{3}{1} \cdot \binom{6-3}{2-1} / \binom{6}{2} = \frac{3}{5},$$

and finally for $i = 2$,

$$p(2) = \binom{3}{2} \cdot \binom{6-3}{2-2} / \binom{6}{2} = \frac{1}{5}.$$

Now, the expected compound yield Y can be calculated (to two significant figures) as

$$Y = \sum_{i=0}^{2} (y(i) \cdot p(i)) = \frac{1}{6} \cdot \frac{1}{5} + \frac{1}{3} \cdot \frac{3}{5} + \frac{1}{2} \cdot \frac{1}{5} = 33.33\,\%.$$

This expected compound yield is not dependent on the wafer map, and if we examine the wafers of Fig. 2.5 the actual yield will be $\frac{1}{3}$ since we know that $i = 1$.

2.2.1.2 Generalizing Yield for Repositories of Arbitrary Size

In the previous section, equations for expected compound yield were created assuming that repository size $m = 1$. Now, we assume that m can be of arbitrary size [63]. Thus, one wafer a will be chosen out of a repository of m wafers to match to a wafer b. It is assumed that each wafer in the repository has the same number of bad dies, but may have differing wafer maps.

The function $y(i)$ is independent of repository size, so only function $p(i)$ must be changed. We expand Eq. 2.5 to $p(i, m)$, which now refers to the chance of exactly i bad dies on wafer a being bonded to bad dies on wafer b given a repository of size m. In a static repository, m will decrease by one for each subsequent wafer until all wafers have been bonded, after which point the repository is refilled. In order to express $p(i, m)$, we first define a function $s(i)$ such that:

$$s(i) = \sum_{j=0}^{i} (p(j)) \tag{2.8}$$

In other words, $s(i)$ is the probability of at most i bad dies between wafers a and b being paired during bonding, as opposed to $p(i)$ which is the probability of *exactly* pairing i bad dies during bonding. We then express $p(i, m)$ as:

$$p(i, m) = \begin{cases} q(i)^m - q(i-1)^m & \text{for } i > 0 \\ q(0)^m & \text{for } i = 0 \end{cases} \tag{2.9}$$

The expected compound yield Y is then expressed as a function of m:

$$Y(m) = \sum_{i=0}^{g_a} y(i) \cdot p(i, m) \tag{2.10}$$

To revisit the two example wafers of Fig. 2.5, we can now calculate yield for any repository size. Take, for example, $m = 5$. With a repository of five wafers, $p(i, m)$ for the three possible values of i is calculated as

$$p(0, 5) = q(0)^5 = 0.2^5 = 0.00032,$$

$$p(1, 5) = q(1)^5 - q(0)^5 = 0.8^5 - 0.2^5 = 0.32736,$$

and

$$p(2, 5) = q(2)^5 - q(1)^5 = 1^5 - 0.8^5 = 0.67232.$$

The expected compound yield then becomes

$$\begin{aligned} Y(5) &= \sum_{i=0}^{2} (y(i) - p(i, 5)) \\ &= \frac{1}{6} \cdot p(0, 5) + \frac{1}{3} \cdot p(1, 5) + \frac{1}{2} \cdot p(2, 5) = 44.73\,\%. \end{aligned}$$

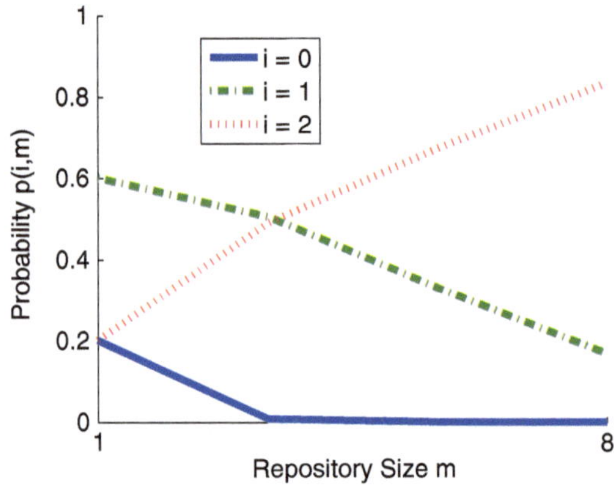

Fig. 2.6 The value of $p(i, m)$ with varying i and m for the two wafer example stack

This is a significant improvement in expected yield over the case of $m = 1$, where yield was calculated as 33.33 %. Indeed, for the two wafers we see significant increases in the probability that a wafer will be matched to another wafer where the locations of more bad dies align. This is shown in Fig. 2.6, which shows the value of $p(i, m)$ for each value of i as m increases from 1 to 8. As m increases, it becomes more likely that all of the bad dies on wafer a will be matched with a bad die on wafer b, as denoted by $p(2, m)$. It is by increasing the likelihood of matching bad dies with bad dies during wafer stacking that we increase expected compound yield by likewise matching more good dies with other good dies.

2.2.1.3 Completing the Recursive Expected Compound Yield Function

The final element to calculate the expected compound yield, while not considering defects created during the stacking process, is to account for n, the number of dies in a stack [63]. Because stacking is an iterative process, the yield calculation must be done recursively. For example, after two wafers are stacked the resulting partial stack is used as wafer b and wafer a is taken from a sorted repository. The partial stack may have more faulty stacks than either of the two previous wafers had faulty dies if there is poor overlap of bad dies in their wafer maps. Therefore, g_b must be recalculated to account for this after each stacking operation. We can say that each wafer at stack level k contains g_k faulty dies.

For a stack of n dies, where n is necessarily greater than one, we can calculate the expected compound yield $Y(n, m)$ starting from a base calculation

$$Y(2, m) = \sum_{i=0}^{g_a} y(i) \cdot p(i, m) \tag{2.11}$$

where $g_a = g_2$ and $g_b = g_1$, just as for the two wafer example. For each subsequent stacking step, $Y(n, m)$ is calculated as

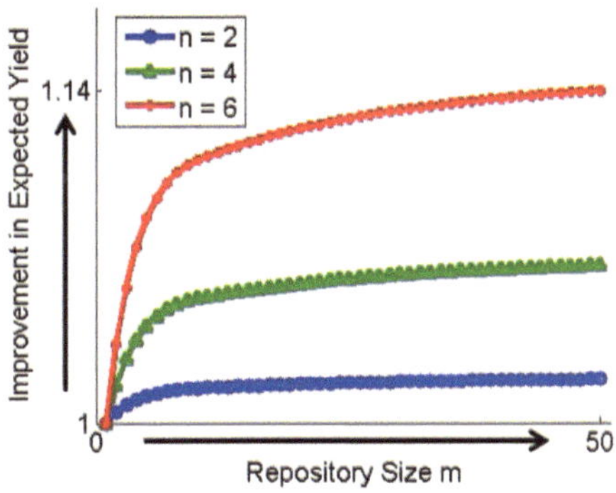

Fig. 2.7 Increase in expected stack yield for various stack sizes n and repository sizes m

$$Y(n,m) = \sum_{i=0}^{g_a} y(i) \cdot p(i,m) \tag{2.12}$$

where $g_a = g_n$ and $g_b = \lfloor (1 - Y(n-1,k) \cdot d + 0.5 \rfloor$.

2.2.2 *Yield Improvements for Repository Matching*

The authors of [63] generated a number of results using a benchmark process with 300 mm wafers. The wafers had an edge clearance (space around the edges of the wafer in which no part of a die may be placed) of 3 mm and a defect density (d_d) of 0.5 defects/cm^2 with a defect clustering parameter (α) of 0.5 [64]. The stack utilized for results consisted of two dies, each of which was a square with an area (A) of 50 mm^2. This results in 1278 dies per wafer with a wafer yield of 81.65 %. Wafer yield is calculated as $(1 + A \cdot d_d/\alpha)^{\alpha}$. From this value, we can calculate g_a and g_b to be 235.

Figure 2.7 shows the increase in expected stack yield for multiple stack and repository sizes. The expected stack yield values are normalized for each stack size. The data presented is derived from the compound yield Eqs. 2.11 and 2.12 without further simulation, so no wafer maps are simulated. As is demonstrated by the figure, a significant yield increase is gained by using even small repositories, with most of the gain in yield seen at repository sizes under 10. As repository size increases further, the gain in yield begins to level off.

The data also reveals that the normalized yield improvement for using repositories is more significant the larger the stack size n. A stack of larger size will have a lower expected compound yield—for the data presented in Fig. 2.7, the expected yield for $m = 1$ was 66.6%, 44.4%, and 29.6% for $n = 2$, $n = 4$, and $n = 6$, respectively. However, with the addition of a repository, the yield improvement was greater for larger stacks. For $m = 50$, the expected stack yields became 67.6%, 47.2 %, and 33.6% for $n = 2$, $n = 4$, and $n = 6$, respectively.

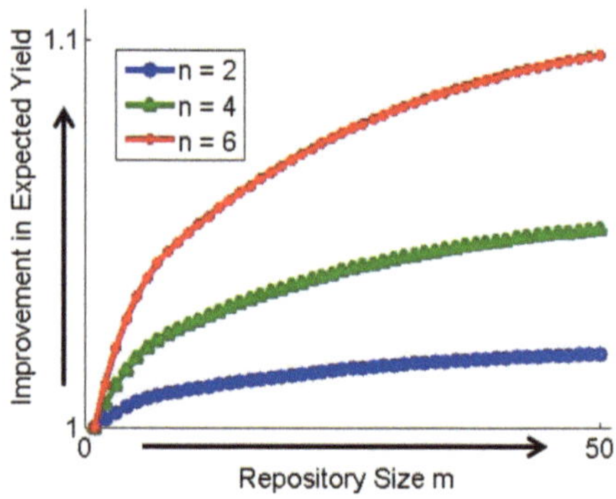

Fig. 2.8 Increase in expected stack yield for various stack sizes n and repository sizes m

In order to evaluate the accuracy of the recursive yield model to a realistic stacking process, the authors of [63] programmed a simulation environment for wafer stacking. The environment consists of a wafer map generator and a wafer matching algorithm for matching and sorting repositories. Wafer maps are generated in such a way that each die on a wafer is generated as good or bad based on a random chance of failure. In this way, wafers have varying numbers and locations of failed dies.

The wafer matching algorithm acts on two static repositories at a time. All wafers in one repository are matched to wafers in the other repository in such a way that all wafers are utilized and the repositories are completely emptied before being refilled. The algorithm is LbL→WbW, so it is iterative in that it matches the repositories for the first two dies in the stack, performs stacking, then matches the partial stack repository to the wafer repository for the next die in the stack, and so on until the stack is complete.

The matching of wafer to wafer between two repositories is an iterative, greedy process. Beginning with two full repositories of size m, each wafer in the first repository is matched against every wafer in the second repository, examining all m^2 combinations of matches, and at the end the match that results in the highest compound yield is chosen. The two matched wafers are removed from the repositories and the algorithm begins anew, now matching $m - 1$ wafers in the first repository against $m - 1$ wafers in the second repository. This greedy process continues until all wafers are matched.

This simulation was performed 10,000 times and the data between simulations averaged to achieve more realistic wafer matching results. Figure 2.8 reproduces Fig. 2.7 using the simulated data as comparison to the mathematical model. As can be seen, the calculated and simulated data produce similar curves, though the normalized expected yield of the simulated data is slightly lower. The actual compound yield for each stack size $n = 2$, $n = 4$, and $n = 6$ for repository size $m = 1$ was 66.6%, 44.4%, and 29.6%, respectively. At $m = 50$, these increased to 67.3%, 46.3% and 32.5%, respectively. These increases of several percentage points for yield are significant, and from the figure it is shown that even small repositories result in substantial yield improvement.

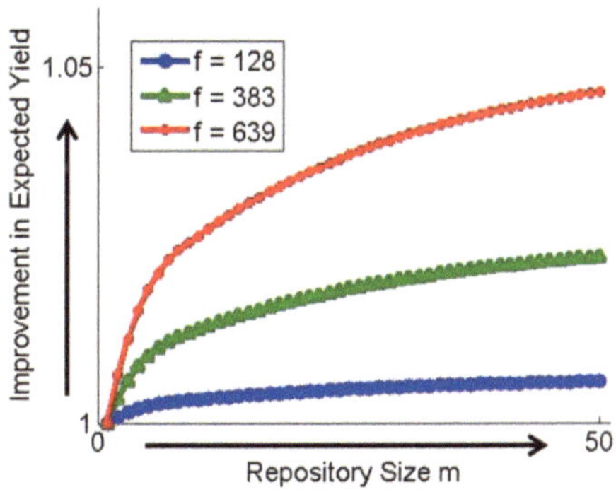

Fig. 2.9 Increase in expected stack yield for various average number of defective dies per wafer f and repository sizes m

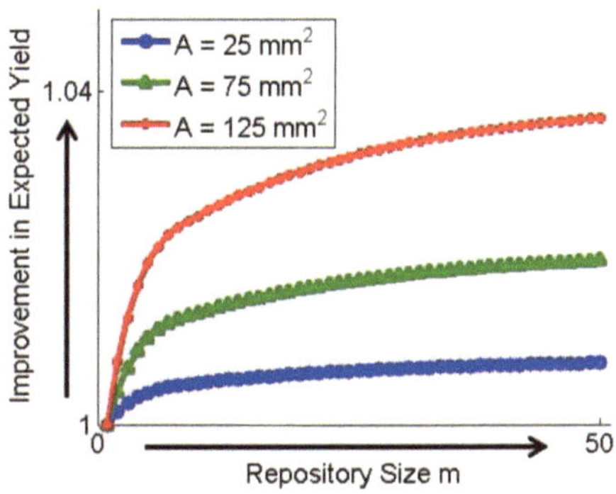

Fig. 2.10 Increase in expected stack yield for dies of varying area size A and repository size m

Figure 2.9 shows the normalized increase in expected compound yield for increasing repository sizes considering three different expected numbers of bad dies per wafer f. Because the wafer can accommodate 1278 dies at the assumed die area of a 50 mm^2 square, each value of f will correspond to an expected yield per wafer. At $f = 128$ expected bad dies per wafer, the expected yield is 90%. At $f = 383$ and $f = 639$, the yields drop to 70% and 50%, respectively. It can be seen that, similar to larger stacks, processes with lower yields per wafer benefit most from the use of repositories. Increasing the repository size for a process with $f = 639$ from $m = 1$ to $m = 50$ results in a change in compound yield from 25.0% to 26.2%. For $f = 383$ and $f = 639$, the change is from 49.0% to 50.0% and from 81.0% to 81.3%, respectively.

Figure 2.10 again shows the increase in expected compound yield for increasing m, but this time for three different die areas A. The number of expected bad dies per wafer remain constant, so increasing the die area will reduce the yield per wafer because the number of dies the wafer can accommodate decreases. It is again seen that the worse the compound yield of a manufacturing process, in this case the process with the largest die area, the greater the benefit from the use of wafer matching. For the worst case die area of 125 mm^2, the absolute yield increases from 44.4 % at $m = 1$ to 46.0 % for $m = 50$. For $A = 75$ mm^2 and $A = 25$ mm^2, this increase is from 57.0 % to 58.1 % and from 80.0 % to 80.3 %, respectively.

The consistent results from all three simulations imply that wafer matching between repositories can provide significant improvement in yield, especially for processes with lower expected compound yield. As the number of dies in a stack, the number of expected bad dies per wafer, or the area of the dies on a wafer increase, the expected compound yield will decrease. Even though wafer sorting can help processes with high m, f, and A, W2W stacking may not be appropriate for these processes. This is because yields are generally very low ($< 40\,\%$) even with repositories. As such, D2W or D2D stacking methods may be utilized to achieve acceptable yields.

2.2.3 Evaluating the Test Cost with Wafer Matching

It is important to evaluate the cost of test versus yield for utilizing wafer repositories [63]. If the overall cost of test is not offset by yield improvement, then repositories would not be a worthwhile investment. In order to use repositories, pre-bond die tests must first be performed to create the wafer map of each wafer. This pre-bond test can significantly increase test cost.

To create a comparative baseline of test cost for W2W stacking, consider a test flow in which dies and TSVs are tested only after stacking. In this case, no pre-bond test is performed, and therefore no wafer repositories can be used. Wafers are blindly stacked and then tested. The cost of this flow can be modeled as follows. Let c_b be the cost of the baseline flow. We introduce the variables c_{die} and c_{TSV}, which are the cost to test each die and the TSV interconnects of each die, respectively. For the purpose of the cost model these values are relative to one another and are of arbitrary units. The variables y_{die} and y_{TSV} take the value of the yield of each die and the interconnects for each stack test, respectively. Using these new variables, we can express c_b as:

$$c_b = d \cdot (n \cdot c_{die} \cdot (n-1) \cdot c_{int}) \tag{2.13}$$

$$+ d \cdot Y(n,1) \cdot y_{die}^{n} \cdot y_{TSV}^{n-1} \cdot (n \cdot c_{die} \cdot (n-1) \cdot c_{int}) \tag{2.14}$$

This cost equation is a sum of products, where each product is the number of items that are tested and the cost of testing those items. In Eq. 2.13, the first product is the test stack (equivalent to d, the number of dies on one wafer) multiplied by the cost for testing each stack, which is the time to test each die ($n \cdot c_{die}$) plus the cost to test the interconnects between each die ($[n-1] \cdot c_{int}$). Since there are interconnects to test only between adjacent dies, the number of interconnect tests are $n-$, recalling that n is the number of dies in the stack.

To factor yield into the cost of test, we only consider performing the final stack test for those stacks that are considered good. This requires factoring in yield, and the number of tests required is $d \cdot Y(n,1) \cdot y_{die}^{n} \cdot y_{TSV}^{n-1}$. The test cost for this final test is the same as for the previous stack test. We do not account for yield loss due to packaging because the comparison test flows should suffer equivalent loss.

Now that we have a baseline model and a method for modeling different test flows, we will define two other flows for comparison. Test Flow 1, whose cost is denoted by c_1, utilizes repositories and wafer matching. This requires pre-bond testing. Furthermore, the stack is tested after stacking and again after assembly and packaging. This results in testing of both of the interconnects after stacking and also a second round of testing for the post-bond dies in addition to pre-bond test. The cost of test for Flow 1 can be modeled as:

$$c_1 = d \cdot (n \cdot c_{die}) \tag{2.15}$$

$$+ d \cdot Y(n,k) \cdot (n \cdot c_{die} \cdot (n-1) \cdot c_{int}) \tag{2.16}$$

$$+ d \cdot Y(n,k) \cdot y_{die}^{n} \cdot y_{TSV}^{n-1} \cdot (n \cdot c_{die} \cdot (n-1) \cdot c_{int}) \tag{2.17}$$

The second test flow we use for comparison, Flow 2, whose cost is denoted by c_2, also utilizes repositories and wafer matching. However, it is an optimized wafer matching test flow that does not retest dies during stacking and instead only tests interconnects. It is assumed that any defects that arise during stacking that would be overlooked by not testing the dies themselves would be caught during package test. The cost of test for Flow 2 is modeled as:

$$c_2 = d \cdot (n \cdot c_{die}) \tag{2.18}$$

$$+ d \cdot Y(n,k) \cdot (n-1) \cdot c_{int} \tag{2.19}$$

$$+ d \cdot Y(n,k) \cdot y_{TSV}^{n-1} \cdot (n \cdot c_{die} \cdot (n-1) \cdot c_{int}) \tag{2.20}$$

Figure 2.11 shows the relative costs of the wafer matching Flow 1 and the optimized matching Flow 2 compared to the baseline test cost c_b. For example, a value of 10% would indicate that the test cost is 10% more than c_b. Figure 2.11a shows relative costs for increasing f, or the number of bad wafers per die, and Fig. 2.11b shows relative cost for increasing stack size. The figure was generated using the values $m = 50$, $c_{die} = 5{,}000$ units, $c_{TSV} = 50$ units^2, $y_{die} = 99\%$, and $y_{int} = 97\%$.

As demonstrated by Fig. 2.11, there is significant increase in test cost for performing pre-bond test on top of post-bond stack and final test. In the worst case, one would expect a roughly 50% increase in test cost, as three test insertions are being performed as opposed to just two. However, it is shown in Fig. 2.11a, b that as the yield of the wafer or the stack decreases (as f or n increases), the cost of test decreases as well. This is because pre-bond test information, for example the wafer maps, can be used to determine which stacks are faulty without performing stack tests. Not performing these extra stack tests cuts cost significantly when yield is low, but as yield increases the added cost of test approaches a 50% increase over c_b.

While Flow 1 always results in an increase in test cost, Flow 2 often results in an overall decrease in test cost. This is because dies are only tested once until final

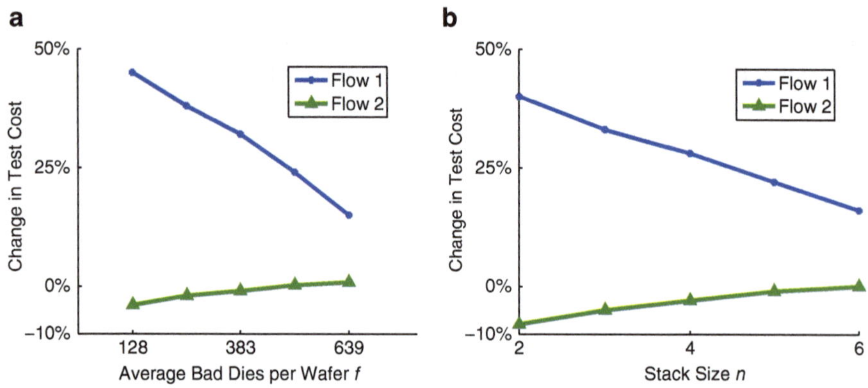

Fig. 2.11 The test cost of matching repositories relative to no wafer sorting for (**a**) a wafer matching only test flow and (**b**) an optimized wafer matching test flow

test so interconnect tests for faulty stacks can be avoided. However, Flow 2 creates a greater onus on test quality for final test, during which time dies that have become faulty during stacking must be detected.

2.2.4 *Summary*

Section Summary

- Static repositories and wafer matching can be utilized to improve stack yield.
- Stack yield decreases with decreasing wafer yield, increasing die size, increasing stack size, and decreasing repository size.
- There are relatively large gains in stack yield with small static repository sizes, although the gains decrease with increasing expected stack yield. The yield improvement drops off as repository size becomes large.
- Test cost increases with the addition of pre-bond test, although this can be offset by reusing pre-bond test data to avoid testing known faulty stacks.

2.3 Yield Benefits of Matching Running Repositories with Different Matching Processes

The authors of [66] conducted experiments considering a variety of different matching processes with running repositories. Consider again that m is the size of each repository, n is the number of dies in a stack, and there are to be e

3D SICs manufactured in a production run. The authors examined three possible WbW→LbL matching processes:

- FIFO1 WbW→LbL
- FIFOn WbW→LbL
- Best-pair (BP) WbW→LbL

The FIFO1 WbW→LbL matching process selects wafers from the repository of wafers that form the lowest tier in the stack in a first-in first-out (FIFO) basis. Because running repositories are used, this repository need not be of greater size than $m = 1$ because the wafer is always removed and replenished. There is also no possible repository pollution for this repository. The wafer chosen from the first repository is then matched to a wafer in a repository for the next wafer in the stack. This process continues until wafers are matched in order from repository 1 to n for the first wafer stack, and then starts again for each subsequent stack. The runtime complexity of the FIFO1 algorithm is $\mathcal{O}(e \cdot m \cdot (n-1)) = \mathcal{O}(e \cdot m \cdot n)$. The worst-case memory complexity is equivalent to the memory needed to store the list of positions of the chosen wafers from each repository, or $\mathcal{O}(n)$.

The FIFOn WbW→LbL matching process attempts to control pollution better than the FIFO1 process. In order to do this, it changes which repository the first wafer is chosen from in a first-in first-out fashion. For example, for a process with $n = 3$ repositories labeled 1 through 3, for the first matching process a wafer is chosen from repository one via FIFO order and matched to wafers in the other repositories. For the second matching process, a wafer is first chosen from repository 2 in FIFO order and matched to the other repositories. After this, a wafer is chosen first from repository 3. After this has been done for all repositories, the sequence starts anew at repository 1. This process can control pollution by forcing the selection of the wafer that has been in a repository the longest when the repository is used to select the first wafer via FIFO order. Thus, a wafer will remain in a repository for at most $n \cdot m$ iterations. The memory and runtime complexities of this process are the same as for the FIFO1 process.

The BP matching process begins by matching a wafer in the first two chosen repositories based on the matching criteria as opposed to any FIFO ordering. This matching continues across the repositories until a complete wafer stack is made. The process then begins anew at the first two repositories until all e stacks are created. The BP process has the greatest freedom in matching wafers, because no wafer choices are forcefully made as with the FIFO process. The drawback is that no effort is made to reduce repository pollution.

The runtime complexity of the BP process is the same as for the FIFO processes. In order to match the first two repositories, m^2 comparisons are made between all the wafers in the first and all the wafers in the second repository. Once the two best-matched wafers are chosen between the repositories, $(n - 2) \cdot m$ more comparisons are needed to match the wafers with the rest of the repositories. For the next iteration of the matching process, only $2 \cdot m - 1$ comparisons are required because the data from the first matching iteration can be reused. This results in a runtime complexity of $(\mathcal{O})(e \cdot m \cdot n + m^2) = (\mathcal{O})(e \cdot m \cdot n)$, which is the same as for the FIFO processes.

The memory complexity is larger than the FIFO processes because, beyond holding the list of positions of the chosen wafers from each repository, m^2 possible match combinations between the first two repositories must also be stored. Therefore, the memory complexity is $\mathcal{O}(m^2 \cdot n)$.

The experimental setup for the simulations in [66] were the same as for the experiments of Sect. 2.2.2. That is, a wafer diameter of 300 mm, wafer edge clearance of 3 mm, a defect density (d_d) of 0.5 defects/cm^2 with a defect clustering parameter (α) of 0.5, square dies with an area (A) of 50 mm^2, 1278 dies per wafer, and a wafer yield of 81.65 %. The production size e was chosen to be 25,000 complete die stacks.

Figure 2.12 shows the increase in the compound stack yield when compared to the yield with repository size $m = 1$ using a FIFO1 WbW→LbL process. Each graph contains curves for three stack sizes of $n = 2$, $n = 4$, and $n = 6$ across repository sizes ranging from 1 to 50. Figure 2.12a shows results using the matching criterion Max(MG), Fig. 2.12b for Max(MF), and Fig. 2.12c for Min(UF). Similar to the results from Sect. 2.2.2, the improvement in yield increased as the expected compound yield of a process decreased, for example as stack size increased from $n = 2$ to $n = 6$. While only data for the FIFO1 process is shown, similar trends were seen for FIFOn and BP processes.

It is also important to note that the Min(UF) matching criterion resulted in the largest yield gains when compared to the other criteria for stack sizes above $n = 3$, but not for smaller stack sizes. Similar results were seen when wafer yield is varied—the Min(UF) matching criterion exceeded the others in yield improvement for wafer yields in the range of 50 % to 70 %, but was worse for larger wafer yields. In the case of large wafer yields above 80 %, Max(MF) outperforms the other criterion.

When comparing FIFO1 to FIFOn, FIFOn always outperforms FIFO1 in terms of yield improvement. This indicates that the FIFO1 process suffers from repository pollution that FIFOn can overcome. The impact of repository pollution on both FIFOn and FIFO1 processes can be seen in Fig. 2.13, which was produced with a fixed repository size of $m = 50$. As the number of stacks that are produced increases, more wafer matching is required during production. As more wafers need to be matched, the yield improvement that can be gained from the FIFOn and FIFO1 processes decreases. This is due to repository pollution decreasing the effective size of the repositories over time. The yield improvement for the FIFO1 process is less effective over time compared to the FIFOn process. This is because the FIFOn process eventually forces polluting wafers out of the repository. As e becomes large, the effect of repository pollution stabilizes. Thus, it is important to utilize a process that can purge polluting wafers over time to maximize yield.

Compared to the two FIFO approaches, the BP process results in the highest compound stack yields. It is also shown in [66] that an adaptive BP process, or one that utilizes a different matching criterion depending on wafer yields, can further improve yield. The BP process utilizing a Min(UF) criterion results in the highest yields when wafer yield is between 50 % and 70 %. This is because good and bad dies arise in similar quantities on a wafer. When wafer yield is lower, the Max(MG)

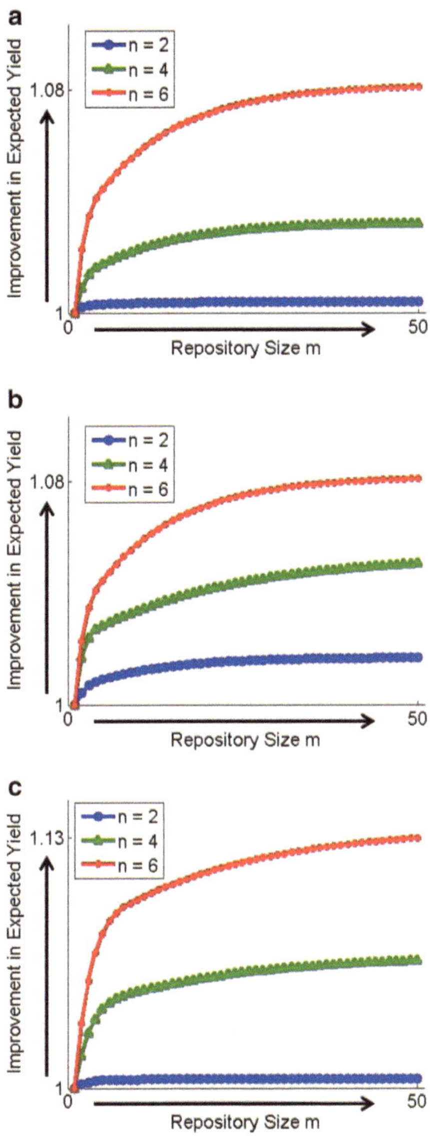

Fig. 2.12 Increase in compound stack yield for various stack sizes n and wafer and running wafer repository sizes m using FIFO1 (**a**) Max(MG) (**b**) Max(MF) (**c**) Min(UF)

criterion tends to produce the highest yield. This is because Max(MG) attempts to match good dies to other good dies, and because good dies are the minority of the dies on the wafer, the best matches produce higher yields. When wafer yield is above 80 %, the Max(MF) criterion produces the highest yields because the bad dies are a minority on the wafer.

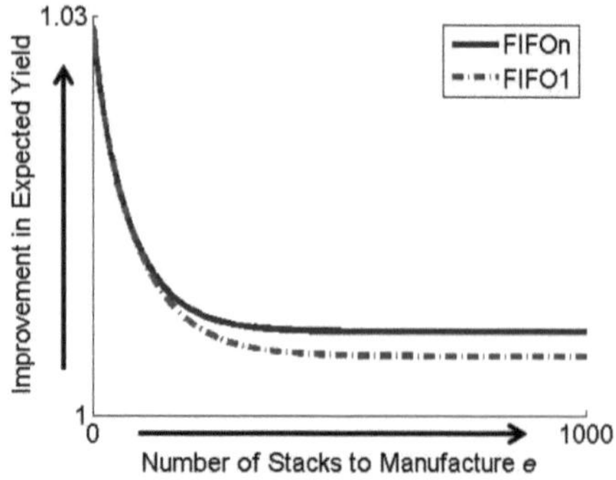

Fig. 2.13 Increase in stack yield for a FIFO1 and FIFOn process as the number of stacks to produce *e* increases

When comparing simulations between running repositories and static repositories, running repositories can yield improvements as high as 2.29 % better than static repositories [66]. Furthermore, the runtime and memory complexity for the running repository algorithms can be significantly better than those for the static repositories, especially for larger n. The test cost for running repositories is similar to those for static repositories shown in Fig. 2.11 and uses similar test flows. However, this does not include the increased complexity and cost of incorporating running repositories into an actual production line.

2.3.1 Summary

Section Summary

- Running repositories can lead to larger improvements to compound stack yield when compared to static repositories.
- Running repository matching algorithms have less runtime and memory complexity than published static repository matching algorithms.
- BP processes result in larger compound yield improvements than FIFOn and FIFO1 processes. Utilizing different matching criterion depending on wafer yields can further improve stack yields.
- FIFO1 and FIFOn processes are both prone to repository pollution as the stack production size increases, effectively reducing compound yield over time.

2.4 Fault Modeling of TSV Resistive-Opens in Stacked DRAM

In a memory stack similar to that shown in Fig. 2.4, TSVs make up the wordlines and bitlines for accessing data in each DRAM cell on a memory array die and route data to and from the peripheral logic on a logic die. Resistive-open defects on these

TSVs result in different faulty behavior when compared with 2D interconnects [67]. This is due to more severe capacitive coupling between the TSV under test and neighboring, densely packed TSVs. This coupling effects behavior based on the type of operation that is occurring on the faulty TSV (read or write) and the voltage of neighboring TSVs.

The authors of [67] studied the impact of resistive-open faults on wordline and bitline TSVs. They performed HSPICE simulations involving a TSV with a coupling capacitance in the range of 0.6–1 fF between the TSV and its adjacent neighboring TSVs. TSVs further away were assumed to have negligible impact. By comparison, the memory cell capacitance was simulated at 30 fF. Wordline capacitances are varied from 1 to 100 fF.

We define notation that will be used for the rest of this section. The notation "Y**w**X" will denote a write operation writing a logic of value X to a memory cell with logic value Y (with X and Y being "1" or "0"). For example, 1**w**0 denotes writing a "0" to a logic cell whose current value is "1". Likewise, the notation "X**r**" denotes a write operation of logic value X. Only read operations and write operations that change the value of the memory cell (e.g. 1**w**0 and 0**w**1) are useful for fault detection.

2.4.1 The Impact of TSV Wordline Resistive-Open Faults

Figure 2.14 shows the schematic view of a resistive-open defect R_F on wordline WL_1 with two parasitic wordlines. The capacitive coupling between WL_1 and the two other wordlines is denoted by C_{1-0} and C_{1-2}. NMOS pass gates are labeled M_0 to M_2 for their respective wordlines. These provide access to the memory cell capacitances $Cell_0$ through $Cell_2$. Bitlines are denoted by BL. Wordline and bitline capacitances, C_w and C_b, respectively, are not shown in the figure.

A large resistive-open defect on WL_1 leaves the gate of pass transistor M_1 floating. The gate-to-source voltage (V_{gs}) of M_1 and the operating region of the transistor will be impacted by a number of factors. The voltage on neighboring wordlines can have an aggressor-victim relationship with the floating wordline and change the voltage at the transistor gate. Second, the drain-to-source voltage (V_{ds}) will change the operating region of the transistor, impacting the charge on the cell. Finally, trapped charges on the floating gate of the transistor left during fabrication will also impact the transistor state.

There are two cases of memory operation that can be affected by the resistive-open on WL_1. In the first case, WL_1 is turned on (goes high) to access $Cell_1$. Both WL_0 and WL_2 will be off because only one wordline per memory bank can be accessed at a time. In this case, the fault can be mapped to traditional AF models for 2D memories because it is a simple stuck-open case. In the second case, either WL_0 or WL_2 is accessed, and this will have a more complex effect on $Cell_1$.

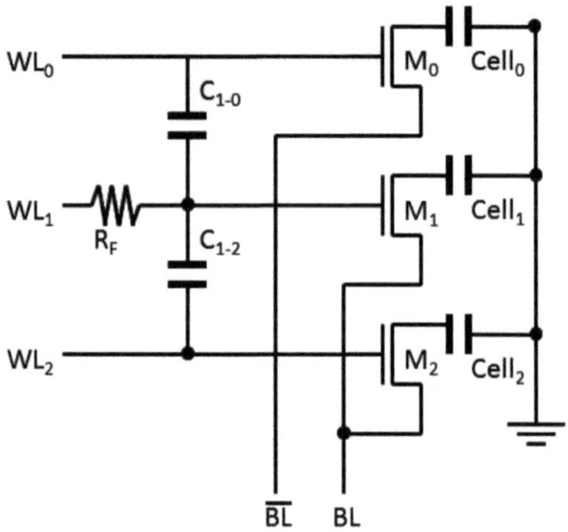

Fig. 2.14 Example of a wordline resistive-open defect with two parasitic wordlines

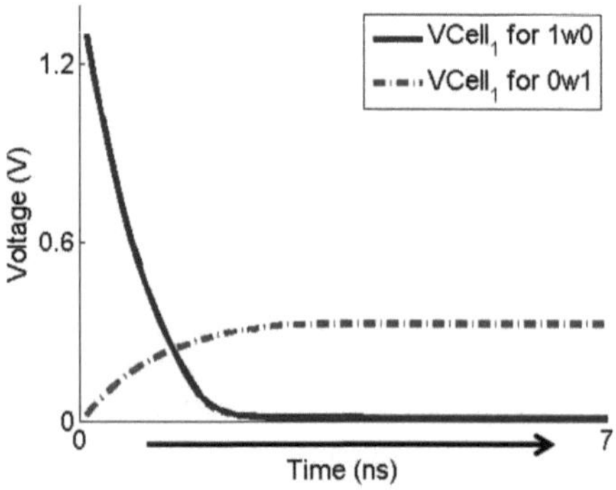

Fig. 2.15 Change in the voltage across $Cell_1$ over time for 1**w**0 and 0**w**1 operations

Returning to Fig. 2.14, when $Cell_0$ is accessed (WL_0 is turned on) $Cell_1$ is floating. These two cells are on complemented bitlines, with the values on the bitlines being fed to a sense amplifier. When $Cell_2$ is accessed, $Cell_1$ is still floating, but they are on the same bitline. There is no functional difference between the two scenarios during write operations, because both bitlines are driven by a source. During read operations there is no source driving the bitlines, and whether or not the accessed cell belongs to the same bitline as the floating cell will impact the faulty behavior of the memory array.

Figure 2.15 shows the change in voltage across $Cell_1$ ($VCell_1$) over time for 1**w**0 and 0**w**1 on a neighboring wordline. The trends shown were similar whether $Cell_0$ or $Cell_2$ were accessed. The wordline capacitance used to generate this figure was $C_w = 10$ fF. As can be seen, a 0**w**1 operation on either neighboring wordline can not pull $VCell_1$ above 0.4 V. This is due in large part to the fact that NMOS transistors exhibit poor pull-up characteristics. For a 1**w**0 operation, the parasitics are significant enough to pull down $VCell_1$. If the capacitive load of the wordline C_w is increased to 50 fF or higher, there is a negligible change in $VCell_1$ for either write operation.

Read operations on neighboring TSVs can also impact $VCell_1$. It is possible that a read operation on a neighboring wordline can cause the values in $Cell_1$ and the accessed cell to read out simultaneously due in part to trapped charges on the gate of the pass transistor creating a high enough V_{gs}. Simulations show that trapped voltages have to be relatively high (0.7 V at minimum) in order to impact read behavior. In this case, $Cell_1$ and a neighboring cell can both be accessed at the same time in a way that impacts read behavior if $Cell_1$ and the neighboring cell are on the same bitline and hold opposite values, or when they are on complementary bitlines and hold the same value.

From these simulations, appropriate fault models can be developed for resistive-open defects on TSV wordlines. In the case of read or write operations that access the defective wordline, the cell associated with the wordline simply cannot be accessed. A write operation cannot charge or discharge the cell capacitance, and a read operation leaves the bitlines at reference voltage. This maps directly to an $AFna$ memory address fault for 2D memories, which is a fault in which a cell is not connected to any memory address. The difference between an $AFna$ fault in 3D memories and 2D memories is that capacitive coupling from nearby blocks or other memory dies may impact the faulty behavior. This is highly dependent on the 3D memory architecture.

Read operations on neighboring wordlines to the faulty wordline can result in reading the value in the accessed cell and the cell associated with the faulty wordline simultaneously. This causes the value on the bitline and the value in the accessed cell to be wrong, which is mapped to a *read disturb fault* (RDF).

2.4.2 The Impact of TSV Bitline Resistive-Open Faults

A resistive-open defect on a bitline can behave differently than a similar defect on a wordline. Consider Fig. 2.16, which shows three bitlines (BL_0 through BL_2) and their complement bitlines, three wordlines of interest (WL_0 through BL_2), and eight memory cells labeled $Cell_0$ through $Cell_8$. $Cell_1$ and $Cell_2$ belong to bitline BL_0, with $Cell_0$ belong to the complemented bitline. This trend continues for the other two bitlines as well. Bitline BL_1 contains a resistive-open fault, and so is left floating.

For any given write operation when a bitline contains a resistive-open defect, whether neighboring bitlines are driven to “1” or “0”, simulations show no significant impact on the value in the memory cells on the faulty bitline. If the cells on the faulty bitline are at 0 V, they will remain at roughly 0 V even if all neighboring bitlines are driven to “1”. Similarly, if the cell contains a logical “1”, there will be no significant reduction in the voltage in the cells on the faulty bitline even if all neighboring bitlines are driven to 0 V. In this way, the bitline open fault for any write operation can be modeled as an $AFna$ memory address fault, similar to read or write access to a wordline with an open fault, because there is no access to the cells associated with the faulty bitline.

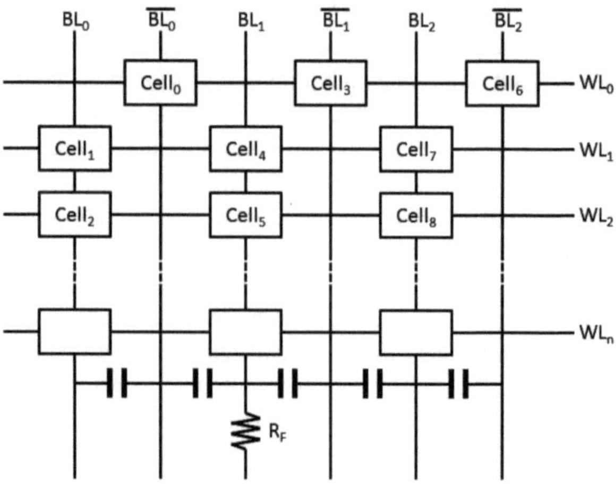

Fig. 2.16 Example of a bitline resistive-open fault in a memory array

The static nature of the value in the cells on the faulty bitline occurs for two reasons. First, the sense amplifier is inactive during write operations such that the cells on the faulty bitline can only discharge via the parasitic capacitance of the faulty bitline. Second, the parasitic capacitance of the bitline is significantly larger than the coupling capacitance of neighboring bitlines, and so is relatively unaffected by values on those bitlines.

The effect of a bitline open fault depends on a read operation is more complex, and depends on which wordline is accessed and on the values stored in memory cells on neighboring bitlines. Since BL_1 is floating, one input to the sense amplifier that compares the bitline to its complement $\bar{BL}_1$ is floating. When wordline WL_0 is accessed, $\bar{BL}_1$ is driven to the voltage stored in $Cell_3$. Even though BL_1 is floating, it will be compared against the value on $\bar{BL}_1$ by the sense amplifier, which will detect a difference and then drive BL_1 and $\bar{BL}_1$ to opposite voltages as though there were no bitline defect. This means that a read operation on WL_0 will not be able to detect a bitline open defect.

If WL_1, WL_2, or any other wordline that is not WL_0 is accessed, the bitline open can cause incorrect memory access behavior. For example, consider accessing WL_1. Because no cell is accessed on $\bar{BL}_1$, both BL_1 and $\bar{BL}_1$ are floating. In order to determine what the sense amplifier will detect as the voltage on BL_1 and $\bar{BL}_1$, the voltages on neighboring bitlines become relevant. BL_1 will experience some coupling with $\bar{BL}_0$, whose voltage will depend on the value in $Cell_1$ on BL_0. Likewise, $\bar{BL}_1$ will experience coupling with BL_2, whose voltage is determined by the value in $Cell_7$. Therefore, the logic values stored in cells on neighboring bitlines becomes relevant in determining the faulty response of the open defect on BL_1.

There are two cases to consider for modeling the faulty behavior of the resistive-open on BL_1 when WL_1 is accessed. In the first case, $Cell_1$ and $Cell_7$ contain the same logic value. Under this circumstance, $\bar{BL}_0$ and BL_1 will be driven to opposite voltages. The coupling between these bitlines, and BL_1 and $\bar{BL}_1$ will cause the voltages in BL_1 and $\bar{BL}_1$ to move toward opposing logical value, as would happen if no resistive-open existed. The difference then becomes amplified by the sense amplifier, driving the faulty bitline and its complement to opposite values. This faulty behavior is termed *compatible coupling* because the logic values in both $Cell_1$ and $Cell_7$ determine the voltages on BL_1 and $\bar{BL}_1$.

The other case to consider is that $Cell_1$ and $Cell_7$ contain opposite logic values. This would cause $\bar{BL}_0$ and BL_1 to be driven to the same voltage, and coupling with BL_1 and $\bar{BL}_1$ will tend to pull these two bitlines to similar voltages as well. To determine the voltage difference between the two bitlines that the sense amplifier will sense, it is necessary to determine whether BL_1 or $\bar{BL}_1$ is more strongly coupled to its neighbor. The further bitlines are from one another, the weaker the coupling between the bitlines. Because $\bar{BL}_1$ and BL_2 are closer to each other than BL_1 and BL_0, the value on $\bar{BL}_1$ will dominate and the sense amplifier will drive BL_1 to its opposite voltage. This is termed *competitive coupling*.

Due to the nature of the read behavior for a bitline open fault, it is modeled as a *neighborhood pattern sensitive fault* (NPSF) and requires three tests to detect on any given bitline. These tests (Tests A, B, and C) are as follows:

$$\text{Test A: } \{\Uparrow (w0); \Uparrow (r0, w1, r1)\}$$

$$\text{Test B: } \{\Uparrow (w1); \Uparrow (r1, w0, r0)\}$$

$$\text{Test C: } \{\Uparrow (w1); \Uparrow (r1)\}$$

2.4.3 Summary

Section Summary

- Due to significantly increased coupling between TSV wordlines and bitlines compared to 2D wordlines and bitlines, 3D memories require different fault models for resistive-open defects on TSVs.
- Accessing a faulty wordline for reading or writing or attempting to write to a faulty bitline results in no cell access, which can be modeled as an $AFna$ fault.
- Attempting to read from a wordline neighboring a faulty wordline can be modeled as an RDF fault.
- When reading from a faulty bitline, the faulty behavior is dependent on coupling from neighboring bitlines and can be modeled as a neighborhood pattern sensitive fault.

2.5 Layer and Inter-Layer Redundancy for Repair of 3D Stacked Memories

A complete view of DfT and yield assurance for 3D memories involves a combination of memory test, redundancy and repair, and wafer matching. The kind of DfT utilized and the types of tests required for 3D memories is depends on the 3D memory architecture, the type of redundancy available for repair, expected stack yields, and the cost of test and repair. It is impossible to review every choice that can be made in these regards in a single section, so instead this section will provide an overview and results for three different memory architectures and their associated repair mechanisms. The first example is a cell array stacked on logic memory architecture that includes entire redundant layers in the 3D stack such that an irreparable die in the stack can be replaced with a new die elsewhere in the stack. The second example is a stacked bank memory architecture that includes a combination of redundancy sharing between memory dies and wafer matching to ensure high stack yields. The last example is a cell array stacked on logic memory architecture where redundant resources, BIST circuits, and BIRA circuits exist only on the bottom die in the stack but are shared across all the memory dies.

2.5.1 Cell Arrays Stacked on Logic with Layer Redundancy

In this example of a repairable 3D memory architecture described in [71], an architecture of cell arrays stacked on logic similar to that of Fig. 2.4 is utilized. Logic cores, cache, and peripheral logic exist only on the lowest die in the stack, and TSVs are utilized as wordlines and bitlines to connect between multiple cell array dies. The dies in the stack are assumed to be independent from each other with respect to redundant resources and fault distribution. For this example, we do not consider the particulars of redundancy on each die. Instead, we consider a die to either be faulty, in which case the redundancy resources were insufficient to completely repair the die, or to be good. The number of good dies in a stack will be termed the die yield Y_D for the stack, and in this example $Y_D = 85\%$.

It is important to consider yield metrics beyond die yield for 3D stacked memories. Manufacturing steps such as thinning and bonding can introduce defects into the dies or the stack that did not exist prior to bonding. We utilize a stacked-die yield Y_{SD} to refer to the number of stacks that are bad due to stacking processes. In this example, $Y_{SD} = 99\%$. A fault of this kind will be assumed to result in a bad memory stack. It is also necessary to consider the yield of TSVs Y_{TSV} in the memory stack. For this example, we will assume that no mechanism exists to repair faulty TSVs, and because the TSVs are utilized by all the memory array dies for memory access, it is assumed that a faulty TSV between two layers causes the entire stacked memory to be considered bad. In this example, $Y_{TSV} = 97\%$.

Using these yield metrics, we can model the effect of layer redundancy on the yield and cost of the 3D stacked memory. Consider a memory stack made up of n non-redundant layers and r additional redundant dies. The total number of layers s of the stack is the number of non-redundant plus the number of redundant layers, or $s = n + r$. For a stack without redundant layers, the total compound stack yield Y can be modeled as a function of the number of dies in the stack and the various yields for the dies, the stack, and the TSVs as follows:

$$Y(n) = Y_D^n \cdot Y_{SD}^{n-1} \cdot Y_{TSV}^{n-1} \tag{2.21}$$

The TSV and stack yields are only considered $n - 1$ times because there are $n - 1$ bonds between dies in a stack of size n.

Adding redundant layers to the stack alters the yield equation. With r redundant layers, there can be r faulty memory array dies in the stack and the stack will still be considered good. If more than r layers are faulty, then the stack must be discarded. In order to determine the yield equation with layer redundancy, we must first calculate the probability $p(i)$ that i out of the s layers in the stack are good. If $i \geq n$, then the stack is good. We define $p(i)$ as follows:

$$p(i) = \binom{s}{i} \cdot Y_D^i \cdot (1 - Y_D)^{s-i} \tag{2.22}$$

The yield for a stack with layer redundancy $Y(n, s)$ can then be written as:

$$\begin{aligned} Y(n,s) &= \left(\sum_{i=n}^{s} p(i)\right) \cdot Y_{SD}^{s-1} \cdot Y_{TSV}^{s-1} = \\ &\left(\sum_{i=n}^{s} \binom{s}{i} \cdot Y_D^i \cdot (1 - Y_D)^{s-i}\right) \cdot Y_{SD}^{s-1} \cdot Y_{TSV}^{s-1} \end{aligned} \tag{2.23}$$

Note that for the case with layer redundancy, there are $s - 1$ bonding steps as opposed to $n - 1$. Both the non-redundant and redundant layers can be faulty, and at least n layers must be fault-free for the stack to be considered good. Equations 2.23 and 2.21 are equivalent when $n = s$, or in other words no layer redundancy is utilized.

If Eq. 2.23 is normalized over Eq. 2.21, then an expression can be developed for the yield improvement of the memory architecture with layer redundancy compared to the architecture without layer redundancy. The yield improvement equation is derived as follows:

$$\begin{aligned} \frac{Y(n,s)}{Y(n)} &= \frac{\sum_{i=n}^{s} p(i)}{Y_D^n} \cdot Y_{SD}^{s-n} \cdot Y_{TSV}^{s-n} = \\ &\left(\sum_{i=n}^{s} \binom{s}{i} \cdot Y_D^{i-n} \cdot (1 - Y_D)^{s-i}\right) \cdot Y_{SD}^{s-n} \cdot Y_{TSV}^{s-n} \end{aligned} \tag{2.24}$$

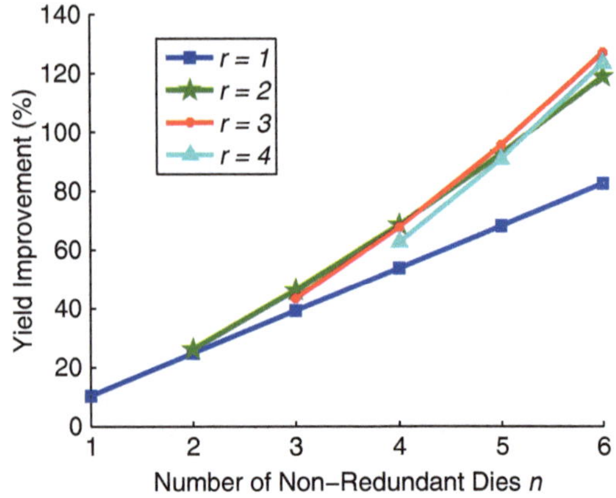

Fig. 2.17 Yield improvement over the base case of no layer redundancy for stacks of varying n and r

Figure 2.17 shows the yield improvement derived from Eq. 2.24 as a percent improvement over the base case of no layer redundancy ($n = s$). The yield improvement is shown for varying n on the x-axis for four values of r. Data for stacks with $r > n$ are not shown, as this would imply that there are more redundant layers than functional layers, which would not be cost effective.

What can be seen from Fig. 2.17 is that layer redundancy improves stack yield for all scenarios with different values for n and r. As the stack becomes larger, the yield improvement also increases significantly due to the increased likelihood of a faulty die existing in the stack. The number of redundant layers that results in the largest yield improvement for a stack is dependent on Y_D, Y_{SD}, and Y_{TSV}. To illustrate this point, consider the case where $n = 3$ and $r = 3$. Despite having more redundant layers than the stack with $r = 2$, $r = 3$ results in a lower yield improvement due to the increased presence of TSV faults and faults caused by stacking. It should be noted that as Y_{SD} increases, the benefit of layer redundancy decreases as the increased stack size begins to have a negative impact on stack yield. For large Y_{SD}, no layer redundancy should be used.

2.5.1.1 Yield and Cost Modeling for Memories with Layer Redundancy

In order to appropriately evaluate the benefits of layer redundancy, the additional cost for manufacturing the stack when redundant layers are added should be considered. Consider C_w to be the cost for manufacturing a wafer and C_{stack} to be the cost associated with all stacking processes for bonding two dies such as TSV creation, thinning, bonding, etc. Then the manufacturing cost $C_m(s)$ can be defined as a function of stack size as:

$$C_m(s) = s \cdot C_w + (s - 1) \cdot C_{stack} \tag{2.25}$$

Note that C_{stack} is only added $s-1$ times to the cost because there are $s-1$ bonds in a stack of size s. Without layer redundancy, Eq. 2.25 can still be used by substituting n for s.

Evaluating the cost benefit of using redundant dies in a 3D memory is dependent on the cost of the stack and the yield improvement for using layer redundancy. This is defined as the cost of a good stack C_{GS} and is defined as the stack manufacturing cost normalized to the stack yield. The cost ratio $C_{GS}(n)$ for a stack without layer redundancy is written as

$$C_{GS}(n) = \frac{C_m(n)}{Y(n)} \tag{2.26}$$

and the cost ratio for a stack with layer redundancy is written as follows:

$$C_{GS}(n,s) = \frac{C_m(s)}{Y(n,s)} \tag{2.27}$$

Dividing Eq. 2.27 by Eq. 2.26 provides a metric for evaluating the cost improvement of using layer redundancy versus no layer redundancy. The resulting equation is written as follows:

$$\frac{C_{GS}(n,s)}{C_{GS}(n)} = \frac{s \cdot \frac{C_w}{C_{stack}} + (s-1)}{n \cdot \frac{C_w}{C_{stack}} + (n-1)} \cdot \frac{Y(n)}{Y(n,s)} \tag{2.28}$$

For a 3D manufacturing cost between 10 % and 90 % of the wafer cost ($0.1 \leq \frac{C_{stack}}{C_w} \leq 0.9$), $\frac{C_w}{C_{stack}}$ has little impact on the cost improvement $\frac{C_{GS}(n,s)}{C_{GS}(n)}$. This is especially true for $n > 3$. Furthermore, it is expected that as n increases, layer redundancy will have a more significant impact on the cost of manufacturing a good memory stack.

Figure 2.18 demonstrates the cost of layer redundancy as a percentage of the cost of a memory stack without layer redundancy (per Eq. 2.28). Results are shown for various values of n and r with $\frac{C_{stack}}{C_w} = 3$. A percentage above 100 indicates that layer redundancy costs more per good stack than producing the same good stack without layer redundancy.

Examining Fig. 2.18, it can be seen that layer redundancy is not cost effective for small stacks. For example, when $n = 1$ and $r = 1$, it is over twice the cost to produce a good stack with layer redundancy than it is to produce a good stack without layer redundancy. This is because adding one redundant layer doubles the cost of manufacturing the stack but does not have a significant impact on stack yield.

As n becomes larger, layer redundancy begins to benefit the good stack cost. At roughly $n \geq 3$, stacks with redundant layers lead to a significant improvement in stack yield that more than makes up for the additional cost of manufacturing a larger stack. It is important to note that the impact layer redundancy has on cost improvement is highly dependent on process parameters, the stack size, and the

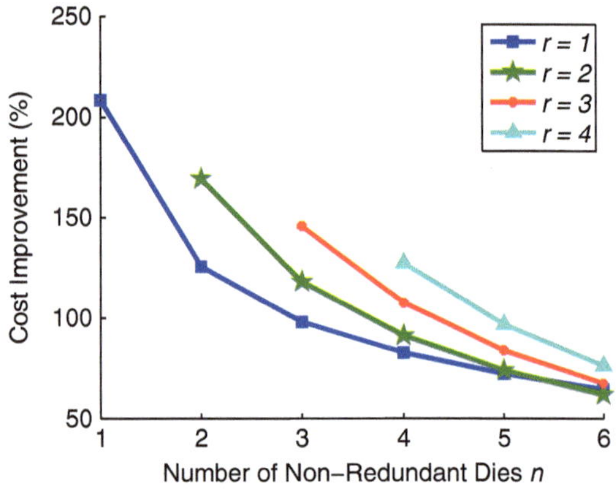

Fig. 2.18 Cost improvement over the base case of no layer redundancy for stacks of varying n and r with $\frac{C_{stack}}{C_w} = 3$

number of redundant layers. For example, as die yield decreases, layer redundancy has an increasingly positive effect on the cost of manufacturing a good stack.

2.5.2 The Impact of Wafer Matching with Redundancy Sharing Between Dies on 3D Memory Yield

In [73], a stacked bank memory architecture is introduced that includes a combination of redundancy sharing between memory dies and wafer matching to ensure high stack yields. The memory stack is built atop a logic die, with each layer having its own peripheral logic and TSVs acting only as a bus for memory data to and from the bottom logic die. Each die contains its own redundant resources for self-repair, but is also capable of sharing its resources with other dies if it does not need to fully utilize them.

Figure 2.19 shows the redundancy sharing architecture used for the shared bank memory used in this section. The image only shows spare rows for simplicity, though spare columns may exist as well. Each die has a programmable decoder for enabling the use of its redundant resources. TSVs link the redundant resources on each die to the decoder on each adjacent die with a multiplexer included that selects between utilizing a resource on the same die or utilizing a resource on an adjacent die. In this way, the memory block on each die not only has access to its redundant resources, but also to the resources of neighboring die. The TSV overhead for the scheme in Fig. 2.19 is $n + m$ TSVs, where n and m are the number of spare rows and columns on each die, respectively.

The benefit of redundancy sharing is that individual memory dies on the stack that have too many faulty rows or columns for self-repair do not necessarily cause the stack to be bad. Consider the two-die stack in Fig. 2.19. If Die 1 has one faulty row and Die 2 has four faulty rows, the entire stack must be discarded without

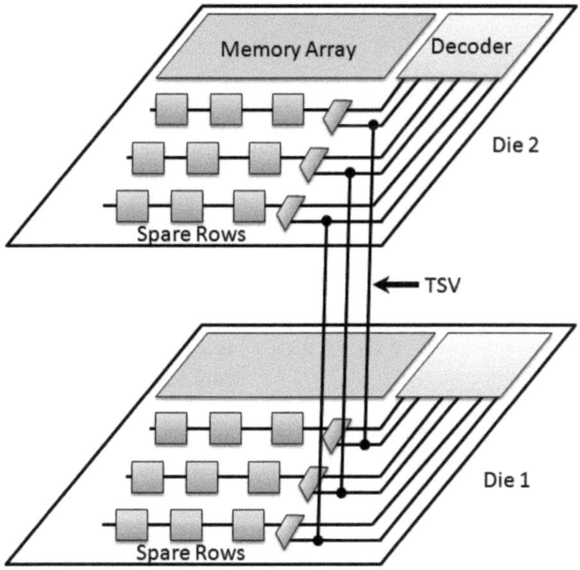

Fig. 2.19 Redundancy resource sharing between dies in a stacked bank memory architecture

redundancy sharing since Die 2 only has three spare rows for self-repair. With redundancy sharing, Die 1 has two extra spare rows that it is not utilizing. Die 2 can utilize one of these extra rows and its own three spare rows in order to completely repair its faulty rows and the stack is considered good.

In [73], the authors utilize an algorithm for wafer matching that considers the redundancy sharing capabilities between adjacent dies to ensure high memory stack yields. The algorithm utilizes a LbL→WbW die matching scheme, because matching more than two dies at any given time while considering redundant resources can be a computationally complex problem. In order to determine if a die is repairable, each faulty bit on a die is classified by whether it can be repaired via spare rows (F_i^r), via spare columns (F_i^c), or via either rows or columns (F_i^o). For a memory block i, let R_i be the number of spare rows and C_i be the number of spare columns on the die. For any two memory blocks i and j on neighboring dies, it is possible to formulate whether the blocks are repairable after stacking if the following equations hold:

$$R_l = (R_i + R_j - F_i^r - F_j^r) \geq 0 \tag{2.29}$$

$$C_l = (C_i + C_j - F_i^c - F_j^c) \geq 0 \tag{2.30}$$

$$R_l + C_l \geq F_i^o + F_j^o \tag{2.31}$$

where R_l/C_l constitutes the number of spare rows and columns remaining after repairing faulty cells that can only be repaired with either a spare row or a spare column. Two memory dies can be bonded together to produce a functioning stack if the above conditions hold for all memory blocks between the two dies. Once the

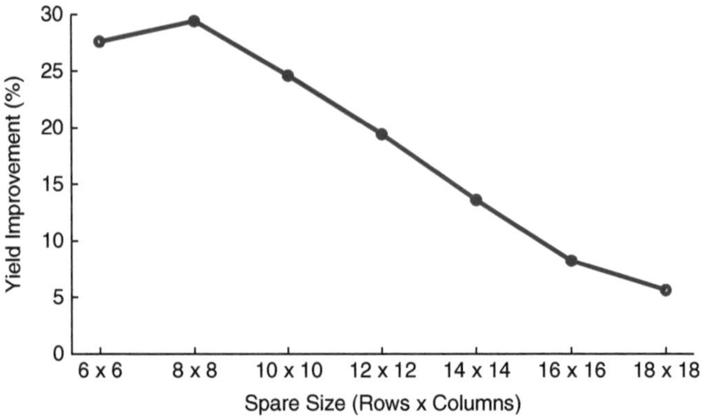

Fig. 2.20 Yield improvement using wafer matching considering redundancy sharing between dies for varying numbers of spare rows and columns

two dies are bonded, the number of spare rows and columns are updated for each block to not include used spares, such that an appropriate repairability analysis can be made with future dies added to the stack.

To assess the yield improvement of redundancy sharing, consider a two-die memory stack. Two static repositories of 500 wafers each are used for matching, where each wafer contains 4×4 memory blocks of 8k $\times$ 8k (Row $\times$ Column) memory cells. Each wafer is subject to a Poisson fault distribution with $\lambda = 2.13$. Among the faults injected via the Poisson distribution, 40% are single cell faults, 4% are double cell faults, 20% are single row faults, 20% are single column faults, 8% are double row faults, and 8% are double column faults. TSV yields between the two dies are 99.9%.

Figure 2.20 demonstrates the percentage yield improvement when using a wafer matching algorithm that considers redundancy sharing between dies when compared to a wafer matching algorithm that matches wafers based only on the ability of a memory block to repair itself using only its spare resources. The wafer matching algorithm used the repairability conditions of Eqs. 2.29–2.31 to determine if adjacent dies would be reparable. Seven different sizes of repair structures were used, ranging from 6×6 spare rows and columns to 18×18 spares.

As can be seen from Fig. 2.20, there is significant yield improvement (nearly 30 % for 8×8 spares) when redundant resources are shared between dies, especially when there are fewer spares on each die. This is because in an environment with fewer redundant resources on each die, there is a greater likelihood that a given die will be unable to completely repair itself using only its own resources. When sharing resources, wafer matching can pair a die that has many faults with a die that has few, resulting in significant yield improvements. As the number of spares increase, the likelihood of a die being capable of repairing all of its faulty rows and columns with its own resources increases, reducing the yield improvement to be gained by redundancy sharing.

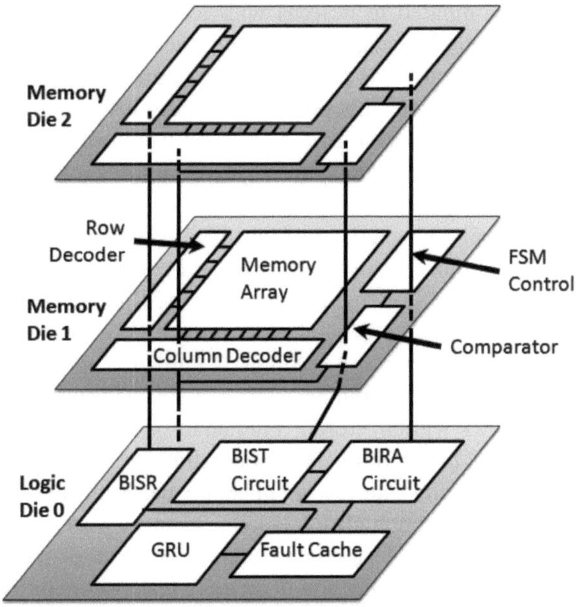

Fig. 2.21 An example block-level schematic of a memory architecture with shared global redundancy and shared test and repair resources on the lowest stack tier

2.5.3 Global BIST, BISR, and Redundancy Sharing from a Single Die in a 3D Memory

The authors of [74] introduce a stacked memory architecture where memory dies share redundant resources, BIST, and BISR circuits that exist only on the bottom logic or memory die in the stack. An example of this architecture is shown in Fig. 2.21 for a 3-die stack. The bottom stack tier is a logic die that would include all logic cores as well as the test and repair circuits and spare resources for the stacked memory dies. These circuits and resources are shared between all memory dies. The memory dies themselves contain banks of SRAM that are distributed across the stacked memory tiers.

The authors use a fault cache-style BISR circuit. The BIST, BISR, fault cache, BIRA, and global redundancy units (GRUs) that contain the redundant resources for all the memory dies are placed on the bottom tier of the stack. Finite state machines (FSMs) and comparators are included on each memory die to interface with the BIRA and BIST circuits to enable testing of all of the memories and to route test data to and from the logic die. As faulty memory cells are located, data from the BIRA is entered into the fault cache and the BISR circuit utilizes this information to distribute redundant resources to the memory dies as necessary to effect repair.

The benefits of a 3D memory with a test architecture similar to that shown in Fig. 2.21 is that the area overhead required for the test and repair circuits is significantly reduced, because each memory die does not need its own circuits. Because redundant resources are globally shared, yield rates are expected to be higher compared to designs with local redundancy because spare resources can be optimally utilized. Furthermore, the test time can be reduced because the test architecture supports parallel testing of all memory dies. A significant drawback to this architecture is that pre-bond memory test cannot be performed because all test circuitry exists on the bottom die, and therefore wafer matching cannot be utilized. This can have a significant negative impact on stack yields unless die yields are relatively high before bonding.

A novel characteristic of the spare resources and repair unit of the architecture in [74] is that, beside spare rows and columns, spare *cylinders* are also included. This is a spare structure that is unique to 3D memories. The spare cylinder is capable of replacing faulty cells along a vertical slice of the memory stack, as opposed to replacing 2D rows and columns. To utilize a spare cylinder, the fault cache stores the local row and column address associated with faulty cells in a vertical swath of the stack without storing a tier address. In this way, an address access on any cell that is part of the vertical column will be redirected to a spare cylinder.

This benefits the test architecture because fault data must be returned to the bottom die through all intervening memory dies via a serial bus, even though the memories can be tested in parallel. If the serial data entering the BISR from higher memory tiers indicates a fault on more than one memory tier, then the BISR does not need to wait for all test data to be returned. A spare cylinder can simply be allocated to the area, reducing test time. Furthermore, spare rows and columns can be saved for instances of faulty cells that would better utilize the spares.

The area overhead of the complete memory architecture is relatively small, which is a benefit of sharing test and repair resources. The additional overhead required by the test and repair architecture comes primarily from two sources. First, each memory die requires its own FSM controller, a comparator, and some additional dedicated logic. The TSVs utilized by these dedicated circuits require further die area. The comparator requires 8 TSVs for a four-die memory stack, the FSM needs $\log_2 4$ TSVs, and the additional logic requires 3 more TSVs for 13 TSVs total. Given a TSV pitch of 10 μm, a 50 nm process, a DRAM density of 27.9 Mb/mm^2, and four memory dies, the area overhead for each die is about 0.46 %, which is relatively small.

For a 3D stacked memory made with the technology given in the previous example with a dies, b memory cells per die, and c bits per memory cell, the area overhead A_h for each memory die can be determined as

$$A_h = \frac{3488 \cdot (\log_2 a + c + 3)}{b \cdot c} \tag{2.32}$$

where $\log_2 a + c + 3$ is the number of TSVs needed by the repair architecture.

The global redundancy architecture also requires additional overhead on the bottom die for the test circuits such as the BIST, BISR, etc. and the GRUs. The fault cache on the bottom die must store all of the address bits for any faulty cell in order to utilize spare resources in any location for faulty memory or rows (this will

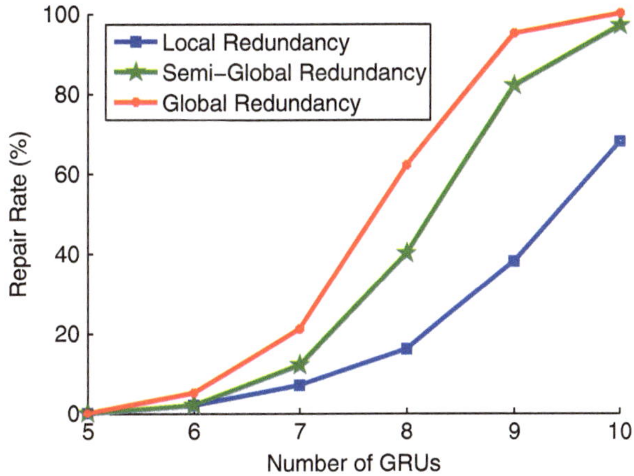

Fig. 2.22 A comparison of the global redundancy architecture with local and semi-global redundancy architectures for varying numbers of GRUs

be discussed in more detail in the next section). In the worst case, the area overhead for the fault cache and all of the test and repair circuits will amount to 11.5 %, though practically it is likely to be less.

2.5.3.1 Allocating Spares and Repair Analysis

In order to allocate redundant resources in 2D memories, an algorithm known as the modified essential spare pivoting (MESP) algorithm [75] is commonly used in industry. The authors of [74] modify this algorithm for the 3D memory architecture of Fig. 2.21. The modified algorithm is referred to as the 3D-Global Essential Spare Pivoting (3D-GESP) algorithm. The 3D-GESP algorithm is different from the MESP algorithm in two primary ways. First, 3D-GESP does not differentiate between spare rows and columns. Each spare entry in the GRU can be utilized as either a row or column as determined by the BIRA circuit. This maximizes use of the spare resources because, for example, the BIRA will never need a spare row when only spare columns remain unused. The second primary difference between 3D-GESP and MESP is that in 3D-GESP each GRU entry can be utilized for any location of faulty memory rows or columns. In MESP, GRU replacements can only be used at the boundaries of memory blocks regardless of faulty cell location.

Simulations were performed to determine the benefits of the global redundant architecture and the 3D-GESP algorithm. A clustered fault model was used for the memory dies with an average of 23.5 faulty cells per die. Each die contained a 1024 $\times$ 1024 $\times$ 8 bit memory array. The number of GRUs on each die is varied and refers to the number of spares available for repair. For example, if there are ten GRUs, then any combination of ten faulty rows and columns can be replaced. The grid size of each GRU is also varied, where the grid size refers to the width of a column or row that each GRU can replace. For example, if the grid size is 32 the GRU can replace 32 $\times$ 8 contiguous bits of a row or column.

Figure 2.22 compares the global redundancy architecture discussed in this section to local and semi-global redundancy architectures. The simulation was performed for 1,000 8-die memory stacks with a GRU grid size of 128. In this context, the repair rate is given as a percentage of the 8-die memory stacks that were completely repairable under each architecture. The semi-global architecture refers to an architecture in which the bottom and top four dies each have access to half of the GRU resources, while for the local redundancy scheme each memory die contains its own share of the GRU resources and does not share them with other dies. The number of GRU resources is varied across the x-axis.

As can be seen in Fig. 2.22, a global redundancy sharing scheme outperforms the semi-global and local sharing schemes across all five GRU sizes. The global repair architecture outperforms the local architecture by 27% on average, with a maximum improvement of 59.9% at eight GRU resources. The global repair scheme outperforms semi-global repair by 8.6% on average, with the largest improvement of 22.3% at eight GRU resources. As the number of GRU resources increases, the improvement in repair rate offered by global repair diminishes, as the local and semi-global repair schemes have more resources available to them to effect complete repair. This shows that the global scheme is better able to distribute spare resources to those dies that require them.

Figure 2.23 demonstrates the benefits of the 3D-aware 3D-GESP algorithm for distributing spare resources compared to the 2D MESP algorithm for the global redundancy memory architecture. Figure 2.23a shows results for a grid size of 4, while Fig. 2.23b, c show results for grid sizes of 8 and 16, respectively. For all grid sizes, the 3D-GESP algorithm significantly outperforms the MESP algorithm in allocating spares to improve stack yield. On average, the 3D-GESP algorithm repairs 8.3 % more memories than the MESP algorithm, with a maximum improvement of 27.6 %.

As the grid size of the spare resources increases, the improvement offered by the 3D-GESP algorithm decreases. This is due in part to the use of a clustered fault model. When the MESP algorithm attempts to repair faulty memories, the need to allocate resources at memory block boundaries requires more GRU entries to effect repair when compared to the 3D-GESP algorithm. However, as grid size increases, the GRU resources are more likely to cover a cluster of memory cell faults with a single GRU resource.

2.5.4 Summary

Section Summary

- Three test and repair architectures have been discussed—layer redundancy, redundancy sharing between dies, and global redundancy sharing.
- Layer redundancy can be a cost effective means to improve stack yield, especially for memory stacks with many dies.

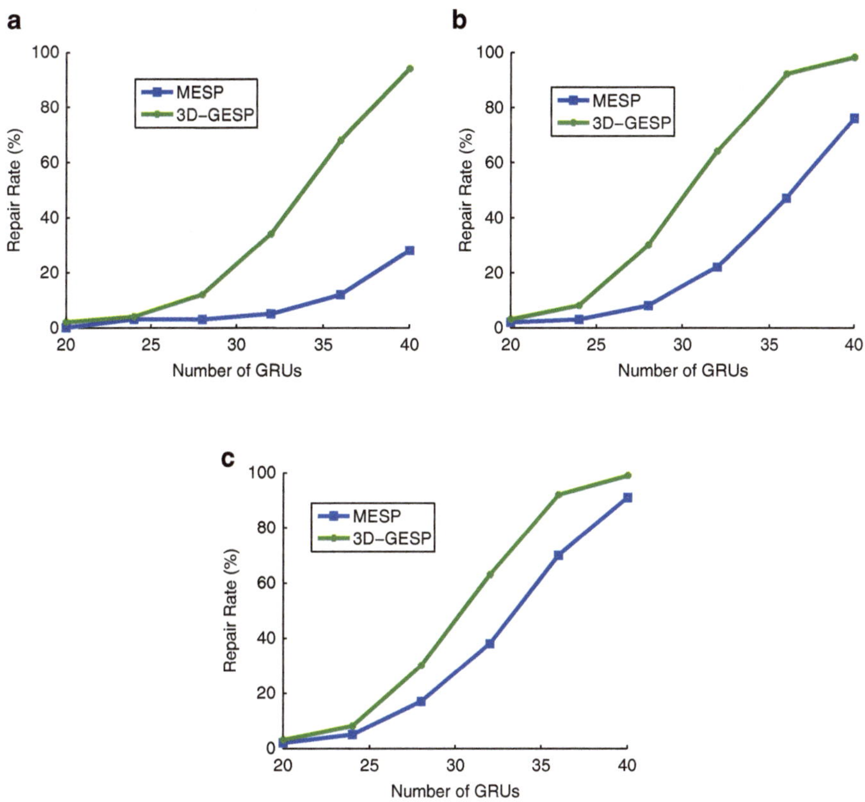

Fig. 2.23 Performance of the 3D-GESP algorithm compared to the MESP algorithm for repair of 3D memories for (**a**) a grid size of 4, (**b**) a grid size of 8, and (**c**) a grid size of 16

- Redundancy sharing between dies, especially when coupled with wafer matching, can significantly improve stack yield. The improvement is more substantial when the number of spares per die is low.
- Global redundancy sharing can decrease area overhead on memory dies and increase yield when compared to semi-global and local architectures, though wafer matching cannot be used as pre-bond memory test is not possible.

2.6 Conclusions

This chapter has addressed an array of topics including wafer matching, 3D memory architectures that can utilize wafer matching and redundancy architectures unique to 3D designs to improve stack yield, and fault models required to detect open defects on TSVs utilized as bitlines and wordlines. If pre-bond KGD test can be made cost effective, then wafer matching can provide significant improvements to stack yield even with small repository sizes. Larger repositories tend to result in better stack yields, though the yield improvement per repository size increase significantly decreases for large repository sizes. Running repositories lead to further improvements over static repositories, although running repositories require more complexity in a manufacturing and assembly flow. Matching wafers via best-pair outperforms other matching processes, and matching criteria also significantly impact yield.

The fault models required to test 3D memory architectures that rely on TSVs as wordlines and bitlines differ from those used to test 2D memories. This is due largely to increased coupling between neighboring TSVs. Accessing a faulty wordline for reading or writing or attempting to write to a faulty bitline results in no cell access. Reading from a wordline neighboring a faulty wordline can result in faulty behavior due to coupling and is modeled as an RDF fault.

Wafer matching and BIST circuits that are aware of the testing needs of 3D memories are important for yield assurance in memory stacks. The type of 3D memory architecture, the kind of redundancy used, and the placement of test and repair circuits and spare resources greatly impact memory yield and cost. Redundant layers can be utilized to improve yield and can be cost effective even though they require significantly higher manufacturing costs. The cost effectiveness of redundant layers increases with stack size and faults per die, although low stacking yields can negatively impact the cost of layer redundancy. Redundancy sharing between layers can be utilized to better distribute spare resources in a stack, and coupled with wafer matching it can significantly improve stack yield. Global redundancy can be used as a low-cost method of test and repair, but cannot be utilized with wafer matching as it does not allow for pre-bond test.

Chapter 3
Built-In Self-Test for TSVs

3.1 Introduction

Pre-bond testing of individual dies prior to stacking is crucial for yield assurance in 3D-SICs [42, 43]. A complete known-good-die (KGD) test requires testing of die logic, power and clock networks, and the TSVs that will interconnect dies after bonding in the stack. This chapter, and Chap. 4 after it, will focus on pre-bond TSV test. In this chapter, we explore built-in self-test (BIST) techniques, their applicability to pre-bond TSV test, and the benefits and drawbacks of BIST approaches.

TSV testing can be separated into two distinct categories—pre-bond and post-bond test [42, 43]. Pre-bond testing allows us to detect defects that are inherent in the manufacture of the TSV itself, such as impurities or voids, while post-bond testing detects faults caused by thinning, alignment, and bonding. Successful pre-bond defect screening can allow defective dies to be discarded before stacking. Because methods to "unbond" die are yet to be realized, even one faulty die will compel us to discard the stacked IC, including all good dies in the stack.

There are further benefits to pre-bond TSV testing beyond discarding faulty dies. Pre-bond testing and diagnosis can facilitate defect localization and repair prior to bonding, for example in a design that includes spare or redundant TSVs. TSV tests that can be performed after microbump deposition can also test for defects that arise in the microbump or at the joint between the TSV and the microbump. Furthermore, dies can be binned for parameters such as power or operating frequency and then matched during stacking.

TSVs play the role of interconnects, hence there are a number of pre-bond defects that can impact chip functionality [38]. Figure 3.1 provides several example defects and their effect on the electrical characteristics of a TSV. Figure 3.1a shows a defect-free TSV modeled as a lumped RC circuit, similar to a wire. The resistance of the TSV depends on its geometry and the material that makes up the TSV pillar. For example, if we assume without loss of generality a TSV made from copper (Cu)

B. Noia and K. Chakrabarty, *Design-for-Test and Test Optimization Techniques for TSV-based 3D Stacked ICs*, DOI 10.1007/978-3-319-02378-6_3,

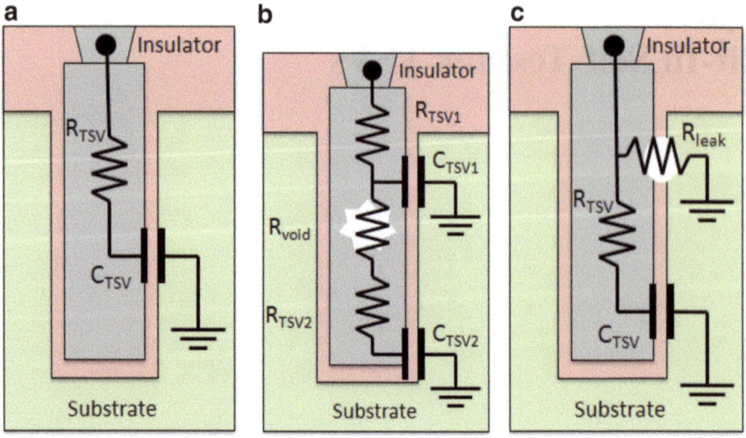

Fig. 3.1 Examples with electrical models for (**a**) a fault-free TSV, (**b**) a TSV with a void defect, and (**c**) a TSV with a pinhole defect

with a diffusion barrier of titanium nitride (TiN), the bulk TSV resistance (R_{TSV}) can be determined as follows [26]:

$$R_{TSV} = \frac{4\rho_{Cu}h}{\pi d^2} \| \frac{\rho_{TiN}h}{\pi\,(d + t_{TiN})\,t_{TiN}} \tag{3.1}$$

where h is the height of the TSV pillar, d is the diameter of the pillar, t_{TiN} is the thickness of the TiN diffusion barrier, and ρ_{Cu} and ρ_{TiN} are the resistivity of copper and titanium nitride, respectively. The resistance of the TSV is determined as the parallel resistance of the resistance of the copper pillar and the resistance of the titanium nitride barrier. Generally speaking, the diameter d of the TSV pillar is much larger than the thickness t_{TiN} of the TiN barrier, and ρ_{TiN} is almost three orders of magnitude larger than ρ_{Cu}. Therefore, R_{TSV} can be approximated as:

$$R_{TSV} \approx \frac{4\rho_{Cu}h}{\pi d^2} \tag{3.2}$$

The capacitance of the TSV can be similarly determined from the pillar and insulator materials and geometries. Once again we assume a copper pillar as well as silicon dioxide (SiO_2) for the dielectric insulator. For the TSV shown in Fig. 3.1a, which is embedded in the substrate with insulator on the sides and bottom of the pillar, the bulk TSV capacitance can be calculated as follows [26]:

$$C_{TSV} = \frac{2\pi\epsilon_{ox}h_c}{\ln[(d + 2t_{ox})\,/d]} + \frac{\pi\epsilon_{ox}d^2}{4t_{ox}} \tag{3.3}$$

where t_{ox} is the thickness of the insulator and ϵ_{ox} is the dielectric constant of SiO_2. The bulk capacitance C_{TSV} is the maximum capacitance under low-frequency, high voltage operation. In Eq. 3.3, the first term on the right models the parallel plate capacitance formed between the TSV pillar and the sidewall. The second term on the right models the capacitance between the TSV pillar and the bottom disk. If the TSV is embedded in the substrate without insulation along the bottom, or after die thinning which exposes the TSV pillar, the TSV capacitance becomes:

$$C_{TSV} = \frac{2\pi\epsilon_{ox}h_c}{\ln[(d + 2t_{ox})/d]} \tag{3.4}$$

Defects in the TSV pillar or the sidewall insulator alter the electrical characteristics of the TSV. Figure 3.1b shows the effect of a microvoid in the TSV pillar. A microvoid is a break in the material of the TSV pillar and can be caused by incomplete fills, stress cracking of the TSV, and other manufacturing complications. These microvoids increase the resistance of the TSV pillar and, depending on the severity of the defect, can manifest as anything from a small-delay defect to a resistive open. In the case of high resistance and open microvoids, the bulk capacitance of the TSV as seen from the circuit end may be reduced as a large portion of the TSV that contributes to the capacitance may be separated from the circuit. This is shown in Fig. 3.1b, as TSV bulk resistance and capacitance are each broken into two separate terms, R_{TSV1} and R_{TSV2}, and C_{TSV1} and C_{TSV2}, respectively.

Figure 3.1c shows a pinhole defect of the sidewall insulator. A pinhole defect refers to a hole or irregularity in the insulator around the TSV that results in a resistive short between the TSV and the substrate. Pinhole defects may be caused by impurities trapped in the insulator, incomplete insulator deposition, stress fractures in the insulator, and more. Depending on the severity of the defect, the leakage of the TSV may significantly increase due to the leakage path to the substrate.

Microvoid, pinhole, and other TSV defects that exist before bonding may be exacerbated over time. Electromigration, thermal, and physical stress can lead to early TSV failures. Many defects cause increased power consumption and heating and can increase physical stress before and after bonding, further aggravating early and long-life failures. A thorough pre-bond KGD test should include a burn-in test for TSVs to screen for these types of failures.

Pre-bond TSV testing is difficult due to a variety of factors. Pre-bond test access is severely limited due to TSV pitch and density. Current probe technology using cantilever or vertical probes requires a minimum pitch of 35 μm, but TSVs have pitches of 4.4 μm and spacings of 0.5 μm [50]. Without the introduction of large probe pads onto a TSV or similar landing pads on the face side of the die [27], current probe technology cannot make contact with individual TSVs. Adding many landing pads is undesirable, as they significantly decrease the pitch and density of TSVs and TSV landing pads. Thus, the number of test I/O available to a tester during pre-bond test is significantly reduced compared to the I/O available during post-bond test. Furthermore, TSVs are single-ended before bonding, meaning that only one side is connected to die logic and the other side is floating (or tied to the substrate in the case of a deposited TSV without bottom insulation). This complicates TSV testing because standard functional and structural test techniques, for example stuck-at and delay testing, cannot be performed on the TSV.

Due to these difficulties, a number of innovative BIST techniques have been introduced in the literature to perform pre-bond TSV test [26, 38, 41, 61]. BIST techniques require few external signals to perform test, often needing little more than power/ground, clock, and BIST enable contacts. Thus, BIST techniques are well suited for the test-access constraints of pre-bond test. Furthermore, they do not require expensive testers or probe cards to interface with the circuit and provide test data or analyze test responses.

BIST techniques also suffer from drawbacks. No current BIST architecture for pre-bond TSV test is capable of detecting resistive defects near or at the end of the TSV furthest from the active device layer. This is due to the fact that these resistive defects result in no significant increase or decrease to the TSV capacitance because the bulk of the TSV closest to the test architecture is intact. Furthermore, BIST techniques do not provide an avenue for TSV burn-in tests, and so cannot screen for TSVs that would experience infant mortality failures. The test circuits utilized by BIST techniques, such as voltage dividers or sense amplifiers, cannot be calibrated before hand and are subject to process variation on the die. This can impact the accuracy of parametric tests.

The rest of this chapter explores several BIST techniques for performing pre-bond TSV test. Section 3.2 explores a technique for detecting TSV shorts based on voltage dividers and comparators that includes a limited repair capability. Section 3.3 examines a DRAM-like TSV test architecture utilizing sense amplifiers. Section 3.4 discusses a test method using ring oscillators along with multiple voltage levels to increase test accuracy. Finally, Sect. 3.5 concludes the chapter.

3.2 TSV Short Detection and Repair Through Voltage Division and Comparators

This section describes a pre-bond BIST architecture [41] that is capable of detecting TSV shorts to ground, which have electrical characteristics as shown in Fig. 3.1c, and binning them into three different categories—*good*, *repairable*, and *bad* TSVs. The categories are differentiated based on the extent of signal degradation on the receiver end of the TSV path.

Figure 3.2 demonstrates some of the effects of TSV shorts. The x-axis of all the graphs is increasing logarithmically with respect to the leakage resistance R_{leak}. Figure 3.2a shows the impact of R_{leak} on signal swing. Leakage resistances of 500 Ω or greater lead to no perceptible change in the voltage of the TSV (V_{TSV}) during a low-to-high transition. Figure 3.2b shows the effect of R_{leak} on delay across

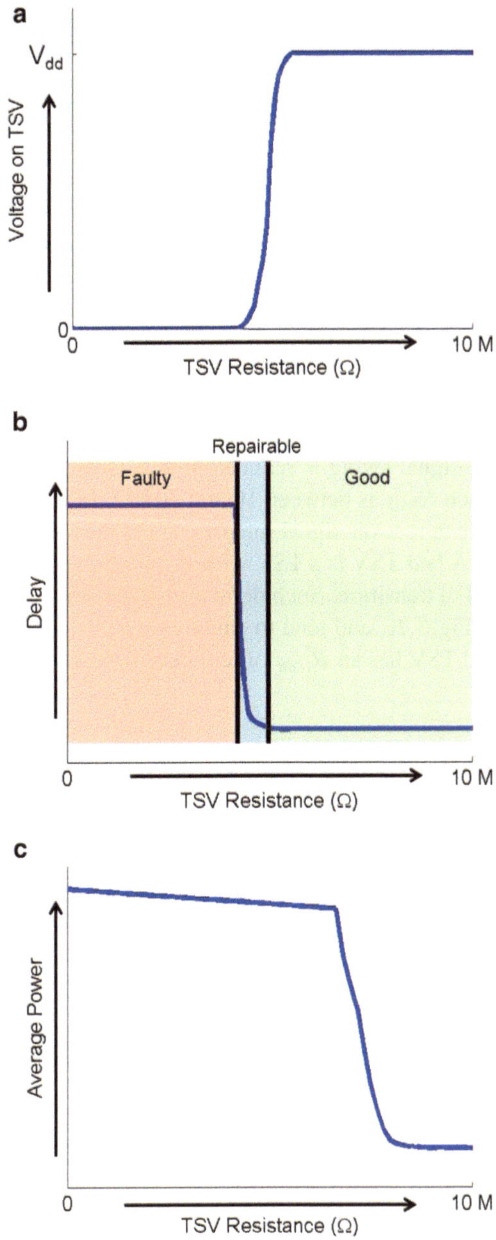

Fig. 3.2 The effect of a TSV short defect of varying magnitude on (**a**) signal swing, (**b**) propagation delay, and (**c**) average driver and receiver power

the TSV. High delays, associated with R_{leak} of around 20 KΩ or less in this example, lead to timing failures on the TSV path. Figure 3.2c shows the impact of R_{leak} on the combined power consumption of the driver/receiver pair across the TSV. High power consumption can lead to power-network and thermal-induced failures.

For the purpose of this architecture, a *good* TSV is a considered to be a TSV across which the voltage after a low-to-high transition event exceeds 90% of V_{DD}. This corresponds to a high R_{leak} resistance, or a TSV pillar whose insulation is intact. In the example of Fig. 3.2a, this occurs when R_{leak} is greater than about 20 KΩ.

A *repairable* TSV is a TSV for which V_{TSV} reaches an acceptable value of V_{DD}, for example above 50%, but is not defect-free. The acceptable value will depend on the range of voltages that will be consistently detected as high by the BIST architecture when considering signal and power supply noise. Such a defective TSV will cause logical transitions at the receiver both before and after repair, although the signal swing is reduced before repair. In the example of Fig. 3.2a, this occurs when R_{leak} is between 10 and 20 KΩ. In this architecture, repairable TSVs may be able to recover signal integrity at the receiver end.

A *bad* TSV is a TSV with a low value of R_{leak}, such that signals across the TSV fail to transition. Such defects also result in significant power consumption, as seen in Fig. 3.2c, and tend to cause excessive heating. Given the example of Fig. 3.2a, a bad TSV has an R_{leak} of less than 10 KΩ.

3.2.1 Design of a TSV-Short Detection/Repair BIST Architecture

Figure 3.3 shows the basic test architecture that utilizes voltage division and a comparator to bin TSV short faults and provide repairability for some TSVs. The TSV is shown on the bottom of the diagram. The TSV pillar in this architecture can be tested if it is embedded in the substrate and completely surrounded by an insulator, or if it is exposed after thinning. This test architecture cannot detect faults on TSVs that are embedded in the substrate without insulator on the bottom of the TSV pillar.

In the architecture of Fig. 3.3, a TSV-Test-Inverter (TTI) is added to the TSV path. The TTI outputs to the net immediately connected to the TSV. During testing of the TSV, the TTI-Gate (TTIG) signal is asserted and the TSV_TEST signal is held low. In this mode, the PMOS resistance (R_{pTTI}) of the TTI and a portion of the TSV pillar resistance forms a voltage divider with the resistance of the TSV short defect (R_{leak}). Because R_{leak} is likely to be much greater than the contribution of R_{TSV} to the voltage divider in a TSV without resistive faults, the voltage at the TTI-TSV network (V_{TSV}) is dependent largely on the value of R_{leak}.

In order to determine the voltage V_{TSV}, the voltage is sampled by a comparator, as shown in Fig. 3.3. It is compared against a reference voltage, V_{ref}, which is chosen to be the minimum voltage that V_{TSV} is allowed to be while considering

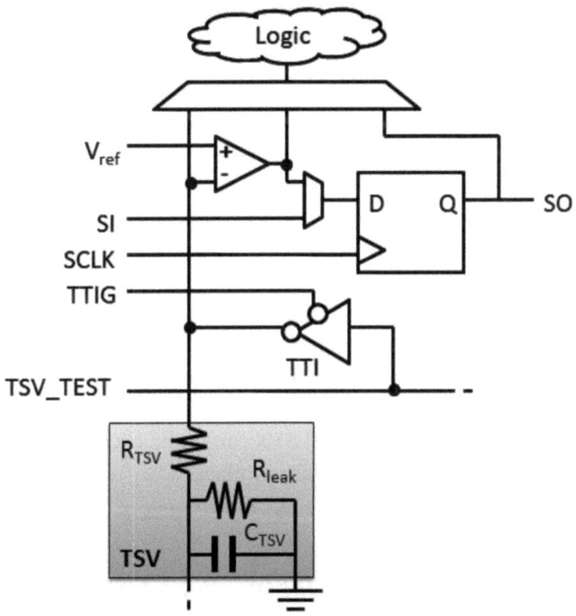

Fig. 3.3 Test architecture for binning TSV short faults with repairability

R_{leak} to be large enough for the TSV to be considered *repairable* or *good*. For example, V_{ref} may be set to 50% or 60% of V_{DD}. Thus, the comparator will output a 1 for a *repairable* or *good* TSV and a 0 for a *bad* TSV.

The output of the comparator can be multiplexed to the input of a scan flop. Figure 3.3 shows the output of the comparator multiplexed with the normal scan input (SI) of the scan flop so that it may be read during test mode. The scan clock (SCLK) is pulsed to latch the comparator output, and because the output is dependent on the value of the voltage divider and requires no complex signal propagation, there is no significant limitation on the timing of SCLK. The output of a comparator is latched into a scan flop across all of the TSVs under test on the die, and the responses are scanned out for analysis.

This first pass of testing separates *repairable* and *good* TSVs from *bad* TSVs, but does not provide the tester with enough information for binning *good* TSVs from *repairable* TSVs. In order to separate the two, a second pass of testing is required. For the second pass, V_{ref} is raised to a higher voltage, for example 90% of V_{DD}, corresponding to the minimum voltage of V_{TSV} that is acceptable for a TSV to be considered *good*. Once again the output of the comparator is latched and the results are scanned out. The results of the second pass can be compared to the results of the first pass. TSVs that resulted in an output of 0 in the first pass are *bad*. TSVs that resulted in an output of 0 for the second pass but not the first are *repairable*. All other TSVs are *good*.

During normal circuit operation or normal test, TTIG is not asserted, creating a high-impedance output to the TTI so as not to interfere with the signal at the TTI-TSV junction. During normal test mode, the scan flop can receive input for standard structural test.

3.2.2 TSV Repair Utilizing the BIST Architecture

The architecture of Fig. 3.3 can be utilized to recover a degraded signal from a *repairable* TSV during normal circuit operation. The comparator, utilizing its first pass V_{ref} value, can act as a level converter. In this case, it receives as an input a degraded signal due to a low R_{leak}, for example 50% or 60% of V_{DD}, and restores it to 90% or higher of V_{DD} at the output of the comparator.

After test, when all *repairable* TSVs are identified, the input signal of the multiplexer between the comparator and the rest of the die logic can be permanently set to pass the output of the comparator, which places the test circuit in *repair mode*. In this way, the comparator recovers any signal passed through the TSV with some small added delay. For *good* TSVs, the multiplexer is set to pass the value at the TTI-TSV junction, which places the test circuit in *bypass mode*. When one or more *bad* TSVs are detected, those TSVs cannot be used and the die must be discarded unless other repair architectures are in place.

3.2.3 Results and Observations for the BIST and Repair Architecture

The feasibility of the proposed architecture was examined through simulations of a 3D SIC benchmark. Simulations were performed on a Fast-Fourier-Transform (FFT) 3D SIC as described in [62] utilizing a 45 nm predictive technology node [56]. The FFT circuit contains 320,000 logic gates spread across two dies in a face-to-back stack. In the design of [62], TSVs were created using a via-first process. Each die in the stack contains six metal layers. Each die has an area of $1.08 \times 1.08\,\text{mm}^2$.

Various designs of the 3DIC were created with different TSV counts. To do so, a min-cut algorithm is used to partition the netlist across two dies. A certain number of cuts are performed, and each cut becomes a TSV on each die. TSVs and standard cells are then placed under constraints of the locations of each TSV. Routing and optimization is then performed to complete the 3DIC. Synopsis Primetime is used for timing analysis.

Utilizing the design in 45 nm and with an appropriate V_{ref}, the architecture is shown to be capable of binning a wide array of R_{leak} values into *good, bad*, and *repairable* categories. Figure 3.4 shows the area overhead of the BIST and repair architecture compared to the area overhead of the TSVs in the design. As the number of TSVs in the design increases, there is a roughly linear increase in both TSV area and the test architecture area overhead. In all cases, the area overhead of the TSVs is much greater than that of the BIST architecture. For example, at 1,500 TSVs the TSVs occupy roughly 20% of the die area, while the test architecture utilizes less than 4% of the die area. At 2,000 TSVs, the area overhead for the BIST architecture remains at less than 5%.

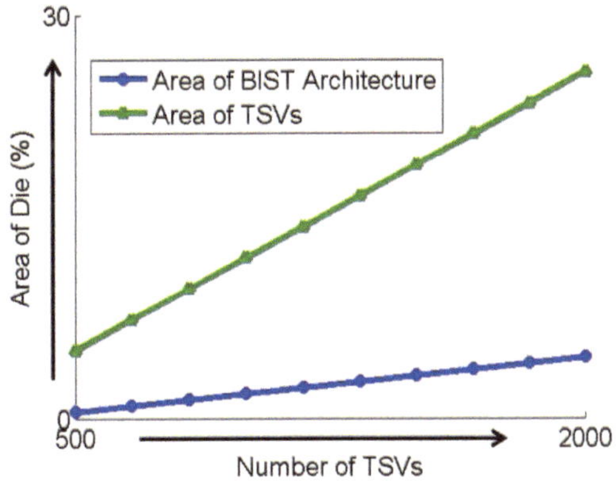

Fig. 3.4 Percent area overhead for TSVs and BIST architecture given the number of TSVs on the die

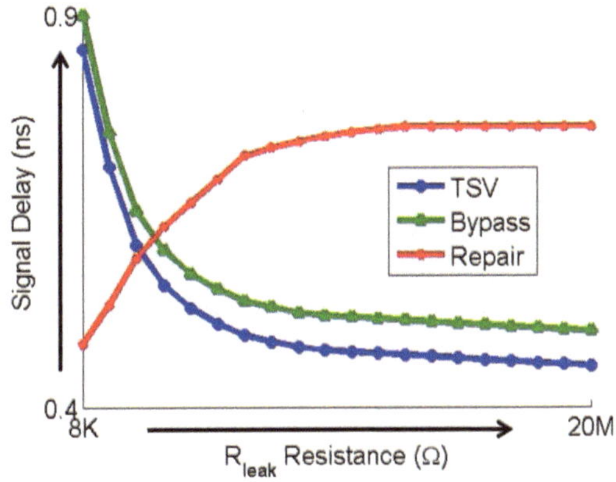

Fig. 3.5 The delay of the TSV path with and without the BIST architecture in repair and bypass modes

Figure 3.5 shows the impact of the test and repair architecture on the delay of the TSV path at various R_{leak} resistances. This delay impact is estimated using HSPICE simulations. Trend lines are provided for the TSV path without the addition of the BIST architecture (TSV), the TSV path with the test architecture in bypass mode (Bypass), and the test architecture in repair mode (Repair). The x-axis is logarithmically increasing. For TSVs with low R_{leak} values, the delay for the TSV and the TSV with the test circuit in bypass mode increases significantly—as high as almost 0.9 ns with an R_{leak} resistance of 8 KΩ. This delay can be significantly reduced using the repair mode of the test architecture when the comparator is used as a level converter. In this case, path delay at an R_{leak} of 8 KΩ drops to less than 5 ns. This is due to the improvement of the signal slew. As R_{leak} increases to values considered *good*, there is shown to be only a small impact on path delay caused by the BIST architecture in bypass mode—roughly 0.3 ns greater than the TSV alone.

3.2.4 Limitations of the BIST and Repair Architecture

The BIST technique is unique among the architectures described in this chapter in that its circuitry can be reused to repair faulty TSVs, which adds significant utility to its use. Furthermore, no digital measurement circuits are used that can contribute to measurement inaccuracy. However, process variation in manufacturing can add significant uncertainty to the BIST architecture described in this chapter. Process variation can result in changes to the PMOS TTI resistance, which can alter the voltage V_{TSV} while the TTI acts as part of the voltage divider resistance. Likewise, process variation in the comparator circuit can further skew binning results.

The severity of TSV faults can also complicate measurement. It is possible for a TSV to exhibit both leakage and resistive defects that would be undetectable by the BIST architecture. Consider, for example, that an R_{leak} of less than $20\,\text{K}\Omega$ is considered faulty. A resistive defect in the TSV of $15\,\text{K}\Omega$ and a R_{leak} of $8\,\text{K}\Omega$ would not register as *bad* using the voltage division method of measurement, even though it is clearly faulty. Such a combination of leakage and resistive changes to a TSV can occur in a number of situations, for example when a large impurity disrupts the TSV pillar and the surrounding insulator.

The *repairable* category of TSVs using this method may have only limited applicability. In a real circuit, a low R_{leak} value is likely to impact TSV reliability over time. The TSV may rapidly degrade over time due to heating, electromigration, etc. While a repair scheme that, for example, replaced a *repairable* TSV with a spare TSV will not be subject to such degradation over time, the repair method of this architecture that utilizes a comparator for level conversion would degrade over time. As the TSV short becomes stronger over time, the repair mechanism of this architecture would eventually fail.

3.2.5 Summary

Section Summary

- An inverter may be added to a TSV path to detect shorts via voltage division.
- A comparator can be utilized at different reference voltages to bin TSVs into *good*, *bad*, and *repairable* categories.
- The comparator can be used as a level converter to offer limited repair to *repairable* TSVs.
- BIST architecture area overhead grows linearly with the number of TSVs.
- Process variation and resistive TSV faults can impact the quality of the test circuit results.
- The repair architecture cannot account for TSV reliability over time.

3.3 DRAM- and ROM-Like Testing of TSVs Using Sense Amplifiers

This section details a pre-bond BIST architecture as described in [26]. Two different test methods are utilized depending on the type of TSV present in the die. A TSV whose pillar is either exposed after thinning or is completely insulated before thinning is known as a *blind* TSV. Blind TSVs can be treated as DRAM cells with their capacitance charged and discharged during test to determine whether the TSV time constant is within an acceptable range. The second type of TSV is called an *open-sleeve* TSV, and in this case the TSV lacks a bottom insulator cap such that the end of the TSV is shorted to the substrate. Open-sleeve TSVs can be treated like NOR-type ROM cells in which a voltage divider is used to determine whether or not a TSV has a faulty resistance.

3.3.1 DRAM-Like Test for Blind TSVs

A DRAM-like test is utilized to determine the time constant of blind TSVs. Groups of TSVs belong to the same test circuit which share a sense amplifier, precharge, and write circuitry. Each TSV within a group is tested sequentially in a five-step process:

1. Reset—All TSVs are precharged to V_{DD}.
2. Hold—TSVs are isolated by turning off access switches. A charge sharing capacitance is biased to a set voltage.
3. Charge share—A charge sharing circuit is created between a TSV and the charge sharing capacitance.
4. Compare to V_{RL}—The charge sharing voltage is compared to V_{RL} via sense amplification.
5. Compare to V_{RH}—The charge sharing voltage is compared to V_{RH} via sense amplification.

The electrical model of the test circuit is shown in Fig. 3.6 for one TSV under test (TUT). The TSV resistance and capacitance are denoted by R_{TSV} and C_{TSV}, respectively. A switch with variable resistance R_{switch} is between the TSV and the

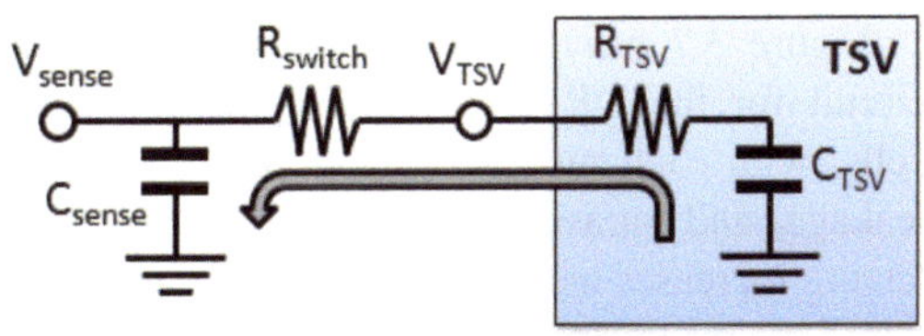

Fig. 3.6 Electrical model of the charge sharing circuit for testing blind TSVs

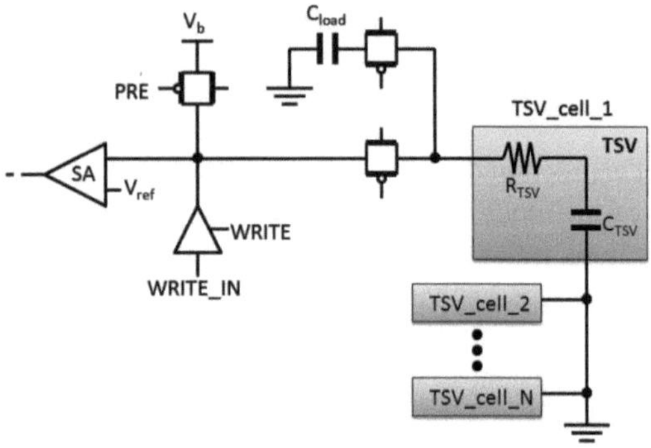

Fig. 3.7 Detailed test module circuit for DRAM-like blind TSV testing

sense amplifier and its corresponding load C_{sense}. The voltages at the TSV (V_{TSV}) and one input of the sense amplifier (V_{sense}) are shown.

Each TSV is effectively treated as a DRAM cell. Switches are added to each TSV cell to allow it to be selectively read and written. Each cell is written to a value such that the charge on the TSV capacitance is known. The value is then read out and compared to a reference voltage via sense amplification.

In the *reset* step, the TSV capacitance C_{TSV} is charged to V_{DD}. This is done simultaneously for all TSVs. During Step 2, or the *hold* step, all TSVs are disconnected from the V_{sense} node by switches, in which case R_{switch} is in a high impedance state. At this point, C_{sense} is charged to a chosen bias voltage, V_b.

In Step 3, or the *charge share* step, the switch corresponding to the TUT is switched to its low-impedance state. This creates a charge sharing circuit between C_{TSV} and C_{sense} in which charge will move from the TSV capacitance to the load capacitance as shown by the arrow in Fig. 3.6. This charge sharing is carried out over a period of time t_{hold} to allow the voltages V_{TSV} and V_{sense} to settle and to account for leakage in defective TSVs. The voltage V_{sense} will thus settle to a different voltage depending on the time constant associated with the TSV.

Steps 4 and 5 perform the sense amplification comparisons between V_{sense} and two reference voltages. Reference voltages are chosen depending on the acceptable bounds of C_{TSV}. A low capacitance, C_l and high capacitance C_h are chosen such that the TSV will be considered not faulty if $C_l \leq C_{TSV} \leq C_h$. Two reference voltages are then chosen, V_{RL} and V_{RH}, to choose acceptable voltages at V_{sense} for a range of time constants. The TSV is considered not faulty if $V_{RL} \leq V_{sense} \leq V_{RH}$. During Step 4, V_{sense} is compared against V_{RL}, and during Step 5 it is compared against V_{RH}.

Figure 3.7 provides a more detailed implementation of a single test module circuit for the DRAM TSV test method. TSVs are attached to the module as cells, with each module having N TSVs. N is chosen based on area overhead, leakage, and parasitic capacitance. As N decreases, the area overhead for TSV testing increases as more test modules and associated circuitry are required to test all of the TSVs in the design. As N increases, the leakage current and parasitic

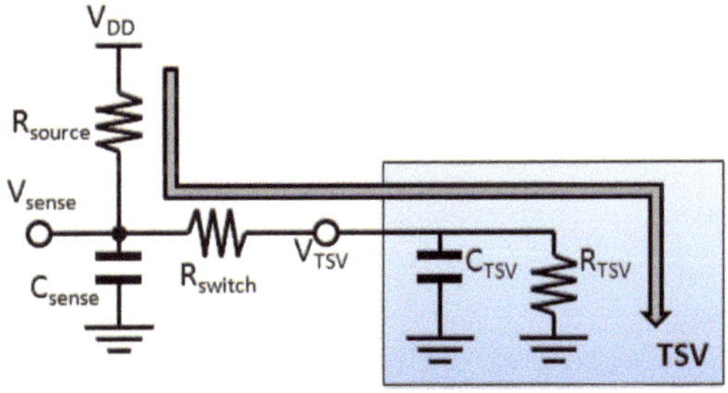

Fig. 3.8 Electrical model of the charge sharing circuit for testing blind TSVs

capacitance increases on the TSV selection line because the number of unselected TSVs increases when an individual TSV is tested.

In the circuit of Fig. 3.7, transmission gates are used to select each TUT as well to bias V_{sense} to V_b during hold. A tri-state buffer is used for the write driver to write to each TSV cell. A sense amplifier is used to compare the voltage after charge sharing to V_{ref}, where V_{ref} is either V_{RL} or V_{RH}.

3.3.2 ROM-Like Test for Open-Sleeve TSVs

A ROM-like test can be used to determine the resistance of open-sleeve TSVs. It is assumed that the end of the TSV, because it is shorted to the substrate, is at ground voltage. As with the DRAM-like test, multiple TSVs can belong to the same test module. In this case, each TSV within a group is tested sequentially in a three-step process:

1. Select—Choose the TUT by opening its switch and closing all other TSV switches.
2. Divide voltage—Create a voltage divider by sourcing a current through the TSV to ground.
3. Sense—Use the sense amplifier to compare the TSV voltage to a chosen voltage level.

The electrical model of the ROM-like test circuit is shown in Fig. 3.8 for one TUT. The TSV resistance and capacitance are denoted by R_{TSV} and C_{TSV}, respectively. A switch with variable resistance R_{switch} is between the TSV and the sense amplifier and its corresponding load C_{sense}. A current source at V_{DD} can be used to source current through the TSV through its on resistance R_{source}. The voltage at the TSV (V_{TSV}) and one input of the sense amplifier (V_{sense}) are shown.

Each TSV is effectively treated as a NOR-type ROM cell. Switches are added to each TSV cell to allow it to be selectively read and written as a column in a memory array. Voltage division is used to create a voltage V_{sense} that is compared to an acceptable voltage through sense amplification.

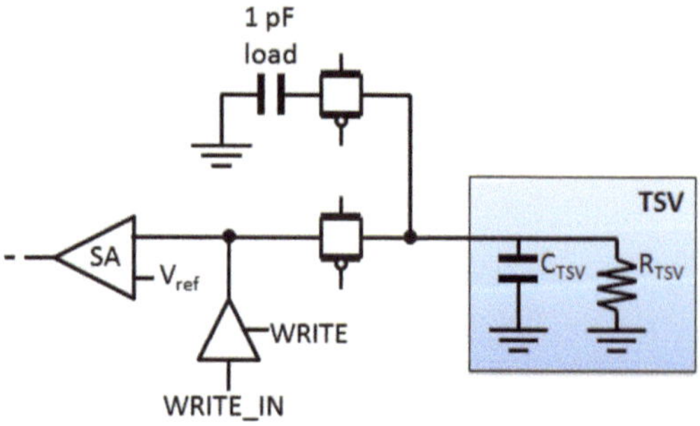

Fig. 3.9 Detailed test module circuit for ROM-like open-sleeve TSV testing

In Step 1, the *select* step, the switch for the TSV that will be tested is opened and all other TSV switches are closed. For the TUT, R_{switch} will be in a low impedance state.

In Step 2, the *divide voltage* step, the current source is connected to the TSV. The resistance R_{source} is the effective source resistance of the current source. This creates a voltage divider as current flows from the current source, through the length of the TSV, and to ground at the open-sleeve end of the TSV. This flow of charge is denoted by the arrow in Fig. 3.8. The value of the voltage at V_{sense} will then be a function of R_{source} and R_{switch}, both of which are known within a process range, and R_{TSV}, which we want to determine.

In Step 3, the *sense* stage of test, the voltage V_{sense} is compared via sense amplification against a reference voltage V_{ref}. The value of V_{ref} is chosen based on an acceptable TSV resistance. A resistance, R_h, is determined such that a TSV is considered not faulty if $0 \leq R_{TSV} \leq R_h$. It should be noted that because R_{TSV} is small compared to the leakage resistance R_{leak}, this test method will not be capable of detecting short defects. The reference voltage is chosen such that, during voltage division, V_{sense} is fault-free if it is less than V_{ref}. For a test module with multiple TSVs, each TSV is tested sequentially by repeating the test steps.

Figure 3.9 provides a detailed circuit implementation of the ROM-like test module. Transmission gates are provided to act as switches to select which TSV is under test and to connect or disconnect the current source. The current source is shown in Fig. 3.9 to have a load of 1 pF. A tri-state buffer is utilized for the write circuitry. A sense amplifier compares V_{sense} to V_{ref}.

3.3.3 Results and Discussion of the DRAM- and ROM- Like BIST

The authors of [26] performed Monte Carlo HSPICE simulations to determine the impact of process variation on the proposed BIST tests. The transistor sizing and control timing is designed to accommodate $N = 100$ TSVs. In other words, each test module should ideally have 100 TSVs belonging to it, and variation from 100 TSVs will produce less accurate test results.

For the blind TSV BIST, the overkill rate, or the rate at which good TSVs are detected as faulty, was shown to be less than 5% for $N = 100$ under process variation. The overkill rate changed significantly with N, being roughly 10% for $N = 50$ and nearly 30% for $N = 200$. Recall that the detection circuit (sense amplifier and associated logic) is calibrated for $N = 100$, and this accounts for a portion of the increase in overkill rate as N deviates from 100. This demonstrates how sensitive the method is to proper calibration and how process variation in the test circuitry may effect test quality.

The escape rate of defects, when a defective TSV is considered good during BIST, was also effected by N and process variation. For all values of N, the escape rate was larger for TSV capacitance values above the highest acceptable capacitance for a fault-free TSV. For $N = 100$, the escape rates were nearly 0% for a 20% increase in TSV capacitance above the largest acceptable fault-free value. For $N = 200$, the escape rate was roughly 5% for a similar 20% increase. As the capacitance of the TSV increases further, escape rates decrease.

Results for the open-sleeve TSV test method imply that the method cannot be used to effectively screen for TSV resistive defects. The overkill rates for both $N = 1$ and $N = 200$ were nearly 50% for nominal values of TSV resistance. If the range of acceptable resistances is increased to a range of about 43 Ω, the overkill rates for $N = 1$ and $N = 200$ become about 9% and 15%, respectively. Escape rates were also high at this relaxed resistance range, at 21% and 18%, respectively.

3.3.4 Limitations of DRAM- and ROM-Like BIST

Although area overhead numbers are not available from published work, the overhead is likely to be high compared to other BIST techniques. The routing difficulty and associated overhead for wiring every TSV in the design to act as memory cells for test purposes alone may be prohibitive. The need for large components such as sense amplifiers and tri-state buffers for each test module is also significant. These components may contain relatively large analog components, and in order to better counteract the effects of process variation, more complex sense amplifiers would have to be used than those that the authors simulated in [41].

As presented in [41], the DRAM- and ROM-like test methods performed poorly under process variation. The high overkill and escape rates, in particular for the voltage-division test circuit, indicate that the BIST method will likely be unusable in an environment with many TSVs. Furthermore, the voltage-division method cannot detect short defects at all. This is not necessarily unexpected, however, as no other BIST method is available for defect detection in open-sleeve TSVs due to the difficulty in testing such TSVs. A better approach would be to test open-sleeve TSVs after die thinning, in which case the TSVs will appear as blind TSVs to any BIST technique.

3.3.5 *Summary*

Section Summary

- Blind TSVs can be treated as DRAM cells and tested via charge sharing and sense amplification.
- Open-sleeve TSVs can be treated as NOR-type ROM cells and tested via voltage division and sense amplification.
- DRAM-like testing resulted in low (5% or less) escape and overkill rates at nominal values of TSVs per test module.
- ROM-like testing is impractical for open-sleeve TSV testing, with very high escape and overkill rates.
- Interconnect routing for testing TSVs as memory cells may be prohibitive in terms of area overhead.
- Depending on the number of test modules required, the area cost of sense amplifiers and write buffers may be large.

3.4 Using Ring Oscillators at Multiple Voltage Levels for TSV Parametric Test

This section describes a method for the pre-bond detection of resistive-open, capacitive, and leakage faults by connecting one or more TSVs to a ring oscillator (RO) [60, 61]. A ring oscillator is a feedback loop consisting of an odd number of inverters. The inversion of the signal causes the output of the RO to oscillate between high and low at a certain frequency according to the equation

$$f_o = \frac{1}{2m\tau_i\tau_p} \tag{3.5}$$

where f_o is the oscillation frequency, m is the (odd) number of inverter stages in the RO, t_i is the delay of an inverter, and t_p is the added effect of the parasitics on all of the interconnects in the RO.

3.4.1 *Ring Oscillator Test Circuit and Defect Model*

Figure 3.10 shows an example of a single TSV connected to a ring oscillator test circuit. The TSV node is flanked by two circuits that make up a bidirectional I/O cell at the TSV. The authors of [61] assume that all TSVs will already have this I/O cell as part of the on-chip circuitry, although in practice (and particularly

Fig. 3.10 TSV and ring oscillator circuit with one TSV

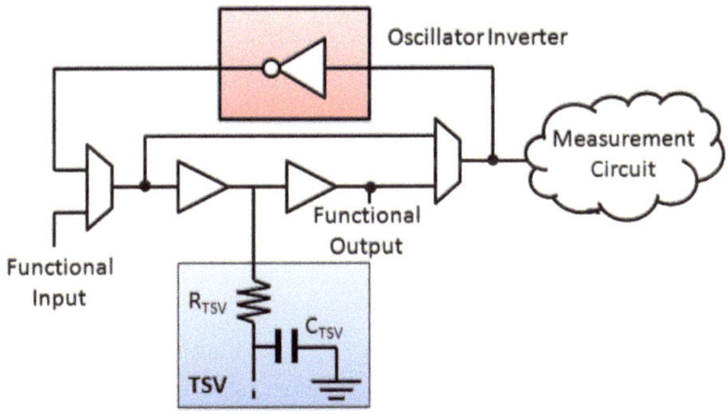

for logic-on-logic dies) the I/O cells may only be unidirectional to save die area depending on functional circuit demands. Around the I/O cell, two multiplexers are added to the design—one to switch between a functional input and an RO input to the TSV, and one to bypass the TSV and I/O cell or to include it in the RO loop. In the figure, a single inverter is utilized to create the RO, and its frequency is measured by a measurement circuit.

In practice, this measurement circuit can be relatively simple, such as a binary counter or linear-feedback shift register (LFSR). The counter or shift register value will change with each pulse of the RO, and after a certain period of time the value in the register can be compared against an expected value. Based on the deviation of the measured value above or below the expected value, an estimate of the oscillation frequency can be made. The binary counter produces a value after a fixed time, and its output can be mapped directly to an oscillation frequency. The LFSR requires fewer gates to achieve the same count limit as the binary counter, but requires a lookup table for its output to determine f_o.

Because the TSV is placed on a node of the RO, changes in the TSV electrical characteristics, for example caused by defects in the TSV pillar or insulator, can impact the oscillation frequency of the RO. As can be seen from Eq. 3.5, the frequency of oscillation changes depending on the parasitics along the RO path. The impact of a TSV defect on f_o will depend on the location, severity, and type of the defect.

Figure 3.11 shows an example of the possible impact of leakage and resistive open faults on f_o. A leakage fault, as shown in Fig. 3.1c, creates a leakage path to ground through some resistance R_{leak}. This leakage causes a driver to charge the TSV capacitance more slowly due to the steady loss of charge. At the same time, the loss of charge leads to a faster discharge of the TSV capacitance. The impact on the charge time of the TSV capacitance is stronger than the effect on its discharge. Thus, over one period of an RO oscillation that includes one low-to-high and one high-to-low transition, f_o will appear longer in the presence of a leakage defect when compared to a fault-free TSV. For a 3 KΩ leakage fault, this effect will increase the propagation delay of the RO signal by about 30 ps.

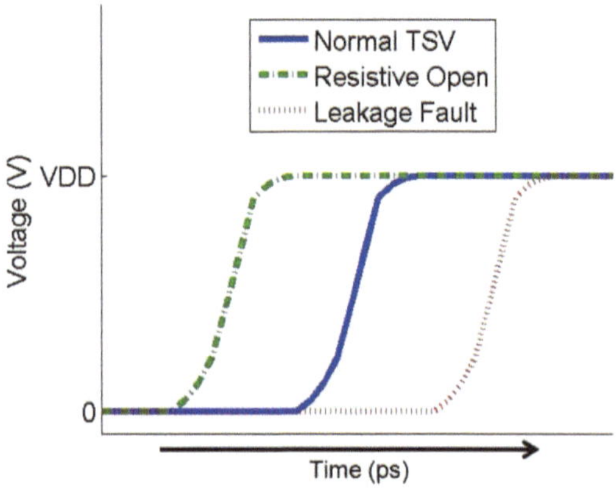

Fig. 3.11 Possible impact of leakage and resistive faults on the voltage waveform of the ring oscillator

A resistive open defect on the TSV, an example of which is shown in Fig. 3.1b, will have a different impact on f_o depending on both its location and severity. In [61], the authors assume that a resistive open defect is abrupt, meaning that the entirety of the resistive increase occurs at a point location along the TSV length. We will keep this assumption for the time being. For an abrupt defect, looking at the TSV from the direction of the I/O cells, the capacitance of the TSV will be split into two capacitances separated by the faulty R_{void} in our example from the figure. The impact of this will be to reduce the effective capacitance of the TSV as experienced by the RO which will in turn reduce the propagation delay of the RO signal as seen in Fig. 3.11.

The closer the defect in the TSV pillar is to the RO node, the more exaggerated the impact will be on f_o because the TSV capacitance will become smaller and smaller. The closer the resistive defect is to the end of the TSV, the less the impact will be on f_o because the capacitance experienced by the RO node will approach the fault-free value of the TSV. Larger R_{void} defects will increase the strength of separation between the TSV capacitances, and thus smaller R_{void} defects will be more difficult to detect regardless of location. To give an example, a 3 KΩ defect halfway in the length of the TSV pillar will reduce the propagation delay of the TSV node of the RO by about 20 ps.

One benefit of using an RO is that multiple TSVs can be strung together as part of the same RO path by duplicating the multiplexers and the bypass path shown in Fig. 3.10 for each additional TSV. The cumulative effect of the parasitics of each TSV on the RO path will determine f_o. In this way, N TSVs can be utilized on the same RO to decrease f_o and reduce the speed at which the measurement circuit would need to operate. The oscillation frequency can be further decreased, or decreased without increasing the number of TSVs on the RO, by adding additional inverters to the RO path. TSVs that belong to the same RO can reuse the same measurement circuit and RO inverters, thereby reducing the area overhead of the complete test architecture.

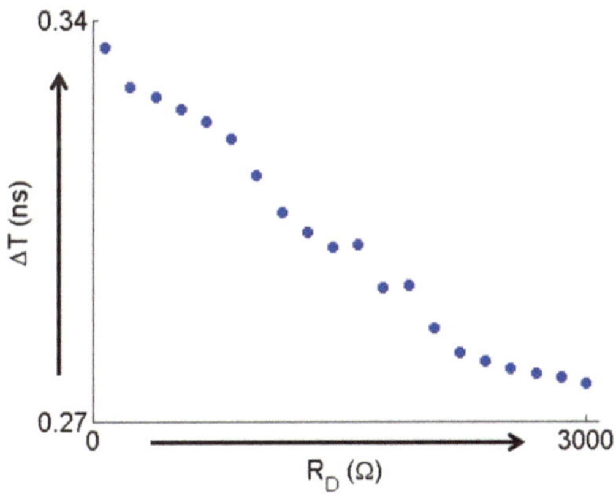

Fig. 3.12 The effect of a resistive open defect halfway in the TSV pillar of varying resistance on ΔT

3.4.2 *Resistive Fault Detection and the Impact of Supply Voltage*

The authors of [61] performed HSPICE simulations for a ring oscillator with $N = 5$. To test for a resistive fault, one of the TSVs in the RO was modeled with an abrupt resistive open halfway in the length of the TSV pillar. In order to determine an accurate measurement of the effect of the TSV parasitics on f_o, an initial measurement of the RO oscillation frequency is taken, and then each TSV is sequentially tested. Alternatively, more than one TSV can be measured at the same time, but at a loss of measurement resolution. For example, to test one TSV, an initial oscillation period T_1 is measured with all TSVs in their bypass state to determine the native RO frequency. Then, an oscillation period T_2 is measured with the TSV under test (TUT) enabled on the RO path and all other TSVs in their bypass state. In each step, T_1 and T_2 are sent to test equipment for evaluation. The important difference ΔT which accounts for the TSV parasitics is determined as:

$$\Delta T = T_2 - T_1 \tag{3.6}$$

By performing this subtraction, the effects of the propagation delay of the I/O cells, multiplexers, and so forth of the TSVs not under test is removed from the final result. Only the delay and parasitics of the TUT are considered. This helps to offset the effects of manufacturing process variations that cause small deviations in gate and interconnect delays that impact the RO frequency.

Figure 3.12 shows the impact of a resistive open defect of varying resistive halfway in the length of the TSV pillar on ΔT. Generally, as the resistance R_D of the defect increases, the impact on ΔT increases while ΔT decreases due to the increasing strength of the separation of the TSV capacitances around the defect. As evident in the figure, small defects of about 500 Ω or less are likely to be undetectable because the faulty signature aliases with the fault-free signature.

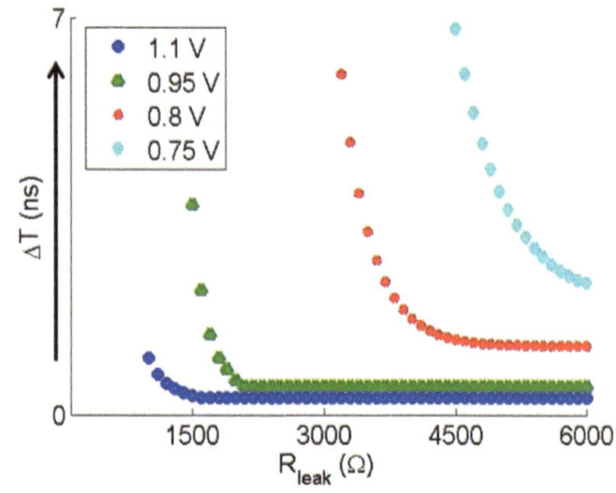

Fig. 3.13 The effect of leakage defects of varying R_{leak} on ΔT at multiple voltage levels

Larger defects are distinctly different from the fault free signature. For example, a defect with $R_D = 1\,\mathrm{K}\Omega$ reduces ΔT by 10% compared to the fault-free case.

The method for detecting resistive defects can perform well under process variations in transistor threshold voltage and gate length appropriate for the simulated technology node. To achieve accurate results under process variation, the supply voltage V_{DD} can be increased. For example, when $V_{DD} = 1\,\mathrm{V}$, there is nearly complete overlap between the faulty and fault-free ΔT values at $R_D = 1\,\mathrm{K}\Omega$. In other words, there is complete aliasing of the faulty and fault-free ΔT. As V_{DD} increases, this aliasing decreases and eventually vanishes at $V_{DD} = 1.25\,\mathrm{V}$. If process variations are more severe, then V_{DD} would have to be increased further to ensure no aliasing. When process variation is small, a higher V_{DD} can be used to increase test resolution and detect smaller resistive open defects or defects toward the end of the TSV pillar.

3.4.3 Leakage Fault Detection and the Impact of Supply Voltage

Leakage faults are detected using the same test flow as described for resistive faults, but have a different effect on ΔT than can be used to tell the two defect types apart. Figure 3.13 shows the effect of the resistance R_{leak} of a leakage defect on ΔT at multiple supply voltage levels. Strong leakage faults with a low R_{leak}, for example less than $1\,\mathrm{K}\Omega$ at $V_{DD} = 1.1\,\mathrm{V}$, result in no oscillation in the RO and appear as stuck-at 0 faults. This is because the I/O cell driver is too weak to overcome the leakage current and pull the TSV node high. This effect is exacerbated when V_{DD} is lower.

If R_{leak} is high enough that oscillation does occur in the RO, it is shown that leakage faults increase the oscillation period, whereas resistance faults decrease the period. Figure 3.13 also demonstrates that at the point where oscillation begins in the

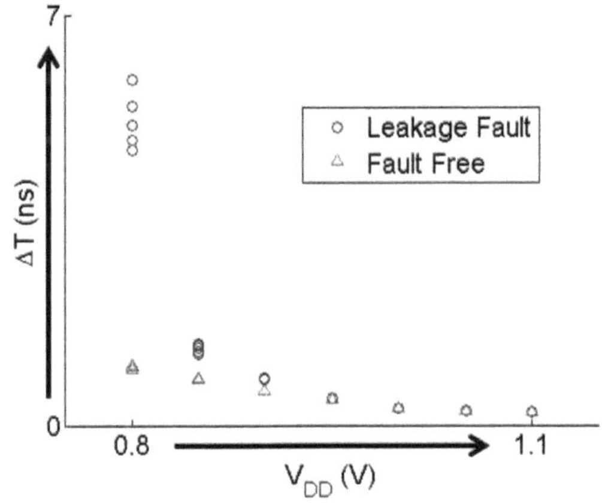

Fig. 3.14 The effect of process variation on the detection of leakage faults at varying V_{DD}

RO, ΔT is significantly dependent on the value of R_{leak}. Therefore, by varying V_{DD} during test, it is possible to detect a wide range of leakage faults. Weaker leakage faults (those with high R_{leak}) will become apparent at lower V_{DD}, and stronger leakage faults will be detectable at higher V_{DD}.

Figure 3.14 demonstrates the impact of process variation on the ability to detect leakage faults at various supply voltages. The figure compares ΔT between a fault-free TSV and a TSV with a leakage fault at $R_{leak} = 3\,\text{k}\Omega$. The high-sensitivity range for a fault of this magnitude occurs around $V_{DD} = 0.8\,\text{V}$. It can be seen that in this range there is no overlap of ΔT between the fault-free and defective TSV. Thus, the defective TSV is detectable in all cases. This difference in ΔT drops off rapidly as V_{DD} increases, such that aliasing under process variation begins starting at $V_{DD} = 0.9\,\text{V}$, with complete aliasing occurring shortly after. This motivates the importance of using multiple voltage levels to detect leakage faults, as the $3\,\text{K}\Omega$ defect would be undetectable at V_{DD} at or above 0.9 V. Stronger leakage defects would require a higher voltage.

3.4.4 Detection Resolution and Area Overhead of the Ring Oscillator Test Circuit

Despite attempts to reduce the impact of process variation on the fault detection resolution of the RO test circuit, there remains some aliasing between ΔT for faulty and fault-free TSVs. In other words, small deviations in the RO oscillation period cannot be distinguished as being caused by process variation or as being caused by TSV defects. The amount of aliasing that will be experienced in an actual circuit depends on the quality of the manufacturing process and the robustness of the circuit layout.

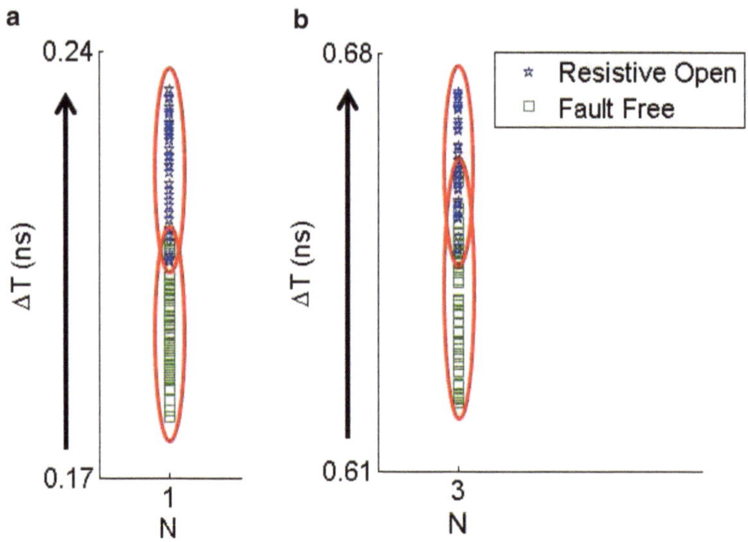

Fig. 3.15 The aliasing effect of testing multiple TSVs simultaneously when (**a**) $N = 1$ and (**b**) $N = 3$

While multiple TSVs can be tested together on the same RO to reduce test time, this testing of multiple TSVs results in a further loss of test resolution. Figure 3.15 demonstrates the aliasing effect of testing multiple TSVs simultaneously on the same RO. The figure was generated under process variation for a 1 KΩ resistive defect in the middle of one of the TSV pillars. Figure 3.15a shows the aliasing between a fault-free and a faulty TSV when the faulty TSV is tested alone. There is only a small overlap between the faulty and fault-free case, therefore aliasing is small and the faulty TSV is likely to be detected. By comparison, Fig. 3.15b shows the effects of testing three TSVs simultaneously when one TSV is faulty. In this case, there is significant overlap between the faulty and fault-free cases, and the faulty TSV may not be distinguishable. This is due in large part to the fact that the effect of process variation is exaggerated when process variations among three TSVs and their associated I/O cells are combined.

There is an inherent measurement inaccuracy to counters and LFSRs that impacts the precision of the RO test method. A clock must be utilized such that a known time interval t is present between the "start" and "stop" signals sent to the measurement circuit. To illustrate this, consider a counter as the measurement circuit. After the "start" signal is sent, the output of the RO is used as the clock signal input to the counter, with each positive or negative edge of the RO output incrementing the counter. After the "stop" signal is sent, the state c of the counter is read out and the oscillation period T^* can be calculated as $T^* = \frac{t}{c}$.

Because the output of the RO will be somewhere between edges when the "start" and "stop" signals are received by the counter, there will be some inaccuracy in the resulting measurement. Consider a positive-edge triggered counter.

The measurement inaccuracy may be up to nearly one RO period in a worst-case scenario, for example when the "start" signal arrives just before one positive edge of the RO output and the "stop" signal arrives just after another positive edge. The upper and lower bounds of the counter state c are thus given by:

$$\frac{t}{T} - 1 \leq c \leq \frac{t}{T} + 1 \tag{3.7}$$

In the worst-case scenario described above, the period T^* as calculated by the external test equipment will be larger than the actual RO period as follows:

$$T^* = \frac{t}{c} = \frac{t}{\frac{t}{T} - 1} = T\left(\frac{t}{t - T}\right) = T + \frac{T^2}{t - T} \tag{3.8}$$

It can be said that the absolute error for the upper bound on T^* is then $E_+ = T^2/(t - T)$. The absolute error E_- for the lower bound, in which T^* is smaller than the actual oscillation period, can be similarly derived to be $E_- = T^2/(t + T)$. We can approximate both errors as $E = T^2/t$, because generally $t \gg T$, so the additional error caused by the inaccuracies of a counter are no worse than $T \pm E$.

The area overhead of the RO BIST architecture can be relatively small due to the reuse of TSV I/O cells. For each TSV on a die, two multiplexers are added as per Fig. 3.10, regardless of whether they share the same RO path or not. Each RO then requires one or more inverters depending on the input requirements of the measurement circuit. These inverters, along with a measurement circuit, are shared among the N TSVs that belong to the same RO path. The authors of [61] demonstrate that the area impact of the multiplexers and inverters of the BIST circuit on a 25 mm^2 die with 1,000 TSVs and $N = 5$ is small (less than 0.04%). In addition, circuitry for test control and the measurement logic for each RO would have to be added to the die.

3.4.5 Limitations of the Ring Oscillator-Based BIST

There are significant benefits to the RO-based BIST when compared to other BIST techniques, including the range of defects that can be detected (resistance, leakage, and likely capacitive) as well as the low area overhead. There still remain, however, significant limitations. Measurement resolution of abrupt resistive open defects can vary widely depending on two conditions. First, it is difficult to detect low resistance (500 Ω or less) defects that increase path delay through the TSV and can lead to reliability issues over the life of the TSV. Even if the defect were situated on the TSV pillar close to the I/O cells, a low resistance may go undetected. Second, due to the reliance of the detection method on changes in the capacitance of the TSV, even large resistive defects may go undetected if they occur toward the end of the TSV pillar.

Actual resistance defect detection may be further obfuscated by the assumption of an ideal and abrupt resistive defect in the TSV pillar. In actuality, voids, cracks, incomplete fills, impurities, and the like may run through the length of a TSV pillar, with the severity of the defect becoming apparent only when integrating along the entirety of the pillar. In such a case, although a resistive defect may be large, its impact on the capacitance of the TSV at any given point can be difficult to detect from the RO period. Other effects may also impact detection resolution, for example jitter in the RO. If multiple TSVs are tested simultaneously, it is possible for the effects of a leakage defect on one TSV and a resistive defect on another TSV to cancel each other out such that the oscillation period of the RO deviates little from the fault-free case.

There exists an inherent trade-off between area cost and test time in the RO BIST architecture. Assuming that TSVs are tested sequentially, test time increases as N increases for the ROs in the test circuit. At the same time, area overhead decreases as fewer RO inverters and measurement circuits are needed. To increase test time, N can be reduced such that more TSVs can be tested in parallel across more ROs, but this in turn increases area overhead. Increasing the number of TSVs tested in parallel on a single RO can also decrease test time, but this comes at a significant decrease in test resolution.

3.4.6 Summary

Section Summary

- Ring oscillators can be used for pre-bond TSV test because their oscillation periods would depend on the parasitics of the TSV under test.
- A counter or linear-feedback shift register can be used on-die to measure the RO oscillation period.
- By varying the supply voltage, defects of wide-ranging severity can be detected in the presence of process variation.
- Area overhead of the BIST architecture can be low due to reuse of the I/O cells that drive a TSV.
- Resistance defect detection resolution can vary significantly with the type and location of the defect along the TSV pillar.
- Inaccuracies in RO period measurement caused by the digital nature of the measurement circuit can contribute to loss of measurement resolution.

3.5 Conclusions

In this chapter, we have examined three distinct BIST architectures for detecting TSV defects before die bonding. An architecture was described that utilized an inverter to create a voltage divider with a TSV in order to detect shorts to the substrate. A comparator is used to determine the severity of the defect and whether or not the TSV can repaired, in which case the comparator doubles as a level converter to ensure the full swing and drive strength of a signal passing through the faulty TSV. Despite this ability to repair, a defective TSV can degrade further over time, eventually failing even with the help of a comparator. Furthermore, process variation can significantly impact the quality of the test circuit.

A second BIST architecture was described that treats TSVs as memory cells for test. A blind TSV can be treated like a DRAM cell, in which it is charged and discharged and its output is directed to a sense amplifier for defect detection. An open-sleeve TSV can be treated like a NOR-type ROM cell, which utilizes a voltage divider whose output is directed to a sense amplifier to detect resistive defects. Testing open-sleeve defects is difficult when using this or any other BIST technique, and so testing of such defects is likely to occur in practice only after die thinning. The complex routing and use of sense amplifiers and write buffers required by this BIST technique may require a large area overhead.

Finally, a BIST technique is described in which one or more TSVs can be tested at the same time as part of a ring oscillator. This technique can detect both resistive and leakage faults and is capable of discerning the difference between the two. Multiple voltage levels must be utilized to ensure the accuracy of defect detection, but when used a wide range of defect severity can be detected under process variation. Despite this fact, resistance fault detection can be difficult when defects are not abrupt or occur toward the end of the TSV pillar. Furthermore, the need for a digital circuit, such as a counter, for detecting the ring oscillator period decreases the accuracy of test.

BIST techniques can offer significant advantages for pre-bond TSV test, especially due to pre-bond test access limitations. BIST requires few external test inputs and outputs, does not require expensive probe cards, does not use complex external testers, and can test all TSVs on a die relatively quickly. On the other hand, many BIST techniques cannot detect every type of defect and do not take into account the need to test microbumps or other features that may be later added to TSVs. Furthermore, all BIST techniques suffer from test accuracy losses caused by process variation in the TSVs and on-die test architectures and no BIST architecture presently designed can detect resistive defects toward the end of the TSV pillar.

Chapter 4
Pre-bond TSV Test Through TSV Probing

4.1 Introduction

Chapter 3 discussed the need for pre-bond TSV test and explored cutting-edge research in TSV testing using BIST. Pre-bond testing allows for the detection of defects that are inherent in the manufacture of the TSV itself, such as impurities or voids, while post-bond testing detects faults caused by thinning, alignment, and bonding. Successful pre-bond defect screening can allow defective dies to be discarded before stacking. Moreover, pre-bond testing and diagnosis can facilitate defect localization and repair prior to bonding. Because methods to "unbond" die are yet to be realized, even one faulty die will compel us to discard the stacked IC, including all good dies in the stack. Pre-bond test is further required if wafer matching, die binning, or other methods for increasing stack yield are to be used, because they require KGD test to be completed for all dies on a wafer.

TSVs play the role of interconnects, hence there are a number of pre-bond defects that can impact chip functionality [38]. Incomplete metal filling or microvoids in the TSV increase resistance and path delay. Partial or complete breaks in the TSV result in a resistive or open path, respectively. Impurities in the TSV may also increase resistance and interconnect delay. Pinhole defects can lead to a leakage path to the substrate, with a corresponding increase in the capacitance between the TSV and the substrate.

Pre-bond testing of TSVs is difficult due in part to TSV pitch and density. Current probe technology using cantilever or vertical probes requires a minimum pitch of 35 μm, but TSVs have pitches of 4.4 μm and spacings of 0.5 μm [50] or smaller. Without the introduction of large probe pads onto a TSV [27], current probe technology cannot make contact with individual TSVs. Furthermore, TSVs are single-ended in the dies at a pre-bond stage, meaning that one end of the TSV is floating or grounded. This complicates BIST because logic can only exist on one end of the TSV before bonding.

The BIST techniques examined in Chap. 3 help to enable pre-bond TSV test by alleviating test access issues, e.g. limited external test access and probe technology

B. Noia and K. Chakrabarty, *Design-for-Test and Test Optimization Techniques for TSV-based 3D Stacked ICs*, DOI 10.1007/978-3-319-02378-6_4,

limitations. A significant downside of these BIST techniques is that they suffer from limitations in terms of observability and the measurements that are feasible. Many BIST techniques cannot detect all types of capacitive and resistive TSV faults, and no BIST technique can detect resistive defects toward the far end of the TSV pillar that is embedded in the substrate. Furthermore, BIST techniques require careful calibration and tuning for accurate parametric measurement, but this is often infeasible, a problem exacerbated by BIST circuits themselves being subject to process variation. Furthermore, BIST techniques can occupy a relatively large die area, especially when considering the thousands of TSVs that are predicted per die [43] and that TSV densities of 10,000/mm^2 or more [42] are currently implementable.

To address the above challenges and offer an alternative to BIST techniques, this chapter presents a new technique for pre-bond TSV testing that is compatible with current probe technology and leverages the on-die scan architecture that is used for post-bond testing. It utilizes many single probe needle tips, each to make contact with multiple TSVs, shorting them together to form a single "TSV network". The proposed approach highlights the relevance of today's emerging test standards and test equipment, and the important role that tester companies can play in 3D SIC testing. Because the proposed method requires probing, it is assumed in this chapter that the die has already been thinned and that it is supported by a rigid platter (carrier) to prevent mechanical damage during probing. During test, the probe needle must be moved once to allow testing of all TSVs in the chip under test. This method also allows for the concurrent testing of many TSVs to reduce overall test time. Furthermore, significantly fewer probe needles are required to test all TSVs, which reduces the cost and complexity of probe equipment.

4.1.1 Probe Equipment and the Difficulty of Pre-bond TSV Probing

A TSV is a metal pillar that extends into a silicon substrate through the active device layer. A "keep out" area where no active devices may be included is therefore associated with each TSV [42]. Prior to wafer thinning, the TSV is embedded in the substrate and is inaccessible to external probing. During thinning, part of the substrate is removed, thereby exposing the TSVs. There are additional considerations that are important when probing thinned wafers. Due to the fragility of a thinned wafer, it needs to be mounted on a carry platter for testing. Probe cards that use low contact forces may also be required. Furthermore, the probe must not touch down too many times during testing, as this may cause damage to both the TSVs and the wafers. Many devices also lack the buffers needed to drive automated test equipment, particularly through TSVs. Thus, probe cards with active circuitry are necessary; this has been articulated recently as being a focus of research at a major tester company [50].

Although interest in 3D-SIC testing has surged in recent years and a number of test and DFT solutions have been proposed in the literature [35–37, 39, 48, 49], pre-bond TSV testing remains a major challenge. Recent efforts have identified some possible solutions for pre-bond TSV testing. A discussion of TSV defects and several methods for pre- and post-bond testing are presented in [38, 42], and a thorough examination of these issues is given in Chap. 3. In [38], 12 different TSV defect types are highlighted, five of which can arise post-bond from errors in alignment, bonding, or stress, while the rest involve defects that arise prior to bonding. Thus, many defects can be targeted at a pre-bond stage. For example, a microvoid in the TSV increases the resistance of the TSV, while a pinhole defect causes leakage between the TSV and the substrate, thus increasing the TSV capacitance. Most of the pre-bond TSV defect types are resistive in nature [38].

In pre-bond probing of TSVs, surface planarity of the TSVs or microbumps impacts the consistency of contact between the probe needles and the TSVs. Therefore, "spring-loaded" probe technologies can facilitate pre-bond probing of TSVs by providing varying degrees of individual control of contacts involving probe needles. Much attention has been devoted to the manufacturing of spring-loaded probe needles [57, 78–80], as surface non-planarity and other issues have made this a desirable technology for achieving good contact during wafer probing. Proposed techniques include membrane probe cards, thermally-actuated probe needles, and probe needles with electrostatic actuators. Furthermore, because non-planarity also impacts TSV connections made during bonding, recent research has explored the planarization of microbumps [81]. This approach can also reduce non-planarity when it is used for the testing of TSVs with microbumps.

It is also important to examine the contacts and contact resistances that may be expected between a probe card and a TSV/microbump. In [76], low-force contacts were made between probe needles and microbumps, and a worst-case contact resistance of 13 Ω was obtained. This is well within a reasonable range to achieve accurate parametric measurements for TSVs. In [54], the effect of probe needle wear on contact resistance is reported. The length of the needle decreased over time as material was worn away by touchdowns, adversely affecting contact force and the quality of contact. However, the reported results show that even with extensive wear after many touchdowns, contact resistances remained below 3 Ω at 30 °C for certain needle heads, with worst-case resistance being no higher than 40 Ω. If these findings are extended to the problem of contacting a TSV network, in which contact quality will be worse for some TSVs, similar changes in contact resistance can be expected.

Novel probe card architectures to enable pre-bond TSV probing are currently being examined in the literature. Cascade Microtech Inc. has introduced a pyramid probe card that has been demonstrated at a 40 μm array pitch [76]. An illustrative example of this probe card is shown in Fig. 4.1a with four probe needles. The needles are deposited on a membrane substrate to allow for individual actuation to compensate for surface non-planarity between TSVs. The needles themselves present a flat square probe head. Form Factor Inc. has introduced the NanoPierce^{TM} contact head for 3D TSV probing, similarly demonstrated at a 40 μm array pitch [77]. An illustrative example of this probe card is shown in Fig. 4.1b. The probe

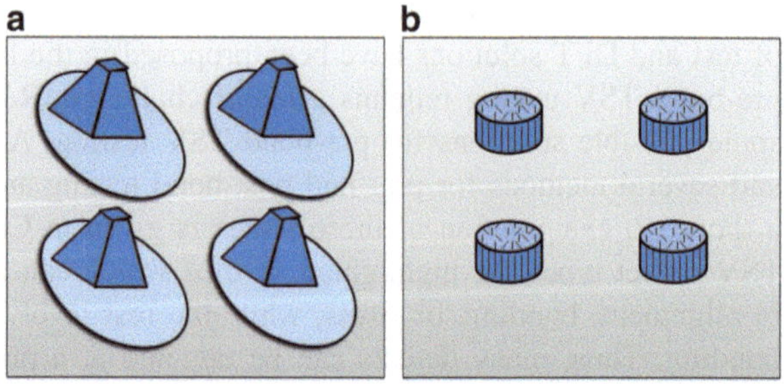

Fig. 4.1 Example of the (**a**) pyramid probe card and (**b**) NanoPierce^{TM} probe card for pre-bond TSV probing

needles are grown from many dense nanofibers that act together to make a contact. Both probe cards utilize low-force probing and show minimal microbump damage.

Despite these new advances in probe card technology, the demonstrated pitches and array placement of needles limit TSV placement and density if individual contact with each TSV required. Furthermore, scaling these probe cards is yet to be demonstrated, and it appears that microbumps may scale faster than probe technology. For example, in [82] microbumps have already been manufactured at sizes of 5 μm with a 10 μm pitch, compared to the 40 μm pitch in [76, 77]. Furthermore, even if every TSV on a die can be contacted individually, the issue of routing test data to and from a probe card with the thousands of probe needles required to contact every TSV is likely prohibitive. If the number of probe needles is instead reduced to a more manageable number, then many touchdowns may be needed to test every TSV on a die, significantly increasing test time and the likelihood of damaging the thinned die or wafer. The probing technique introduced in this chapter ensures that probe technology, regardless of size and pitch, will be capable of testing TSVs with or without microbumps.

The rest of this chapter examines a novel combination of on-die architectures, pin electronic architectures, and test optimizations to enable fast pre-bond parametric TSV test. Section 4.2 introduces a cutting-edge test architecture for pre-bond probing and testing of TSVs. It discusses the benefits and drawbacks of TSV probing and demonstrates the effectiveness of the pre-bond probing architecture. Section 4.3 presents a heuristic approach to reducing pre-bond TSV test time by organizing TSVs into test groups in which multiple TSVs within the same TSV network are tested simultaneously. A examination of the considerable reduction in test time possible through parallel TSV test is provided. Finally, Sect. 4.4 concludes this chapter.

4.2 Pre-bond TSV Testing

In the pre-bond test method discussed in this chapter, a number of TSVs are shorted together through contact with a probe needle to form a network of TSVs. The network capacitance can be tested through an active driver in the probe needle itself, and then the resistance of each TSV can be determined by asserting each

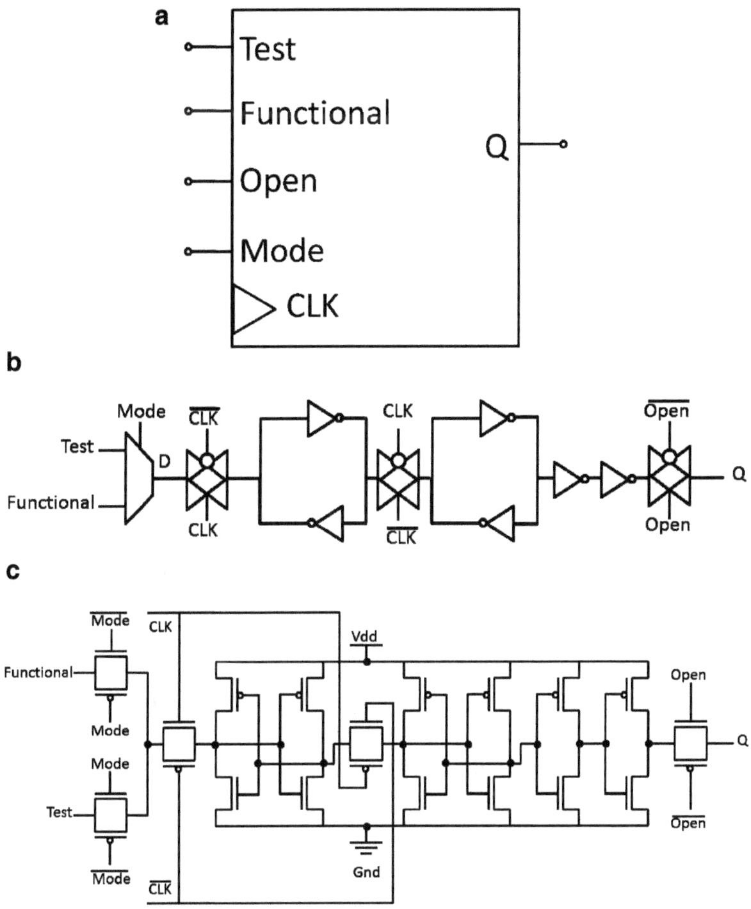

Fig. 4.2 Design of a gated scan flop: (**a**) block-level; (**b**) gate-level; (**c**) transistor-level

TSV onto the shorted net. More details on TSV testing are given in Sect. 4.2.1. In this section, the new test architectures used for pre-bond TSV testing will be discussed.

For post-bond external tests, a 1,500-style die wrapper with scan-based TSV tests has been proposed [35] and is discussed in detail in Chap. 7. It is assumed in this section that die-level wrappers are present. To enable pre-bond TSV probing, the standard scan flops that make up the die boundary registers between die logic and TSVs are modified to be gated scan flops (GSFs), as shown in Fig. 4.2. Alternative methods to accessing a TSV on a die have been reported as in [45], but these methods are not compatible with die wrappers, and so are not considered in this chapter as die wrappers are likely to be utilized in future 3D stacks. As seen at the block level in Fig. 4.2a, the gated scan flop accepts either a functional input or a test input from the scan chain; the selection is made depending on operational mode.

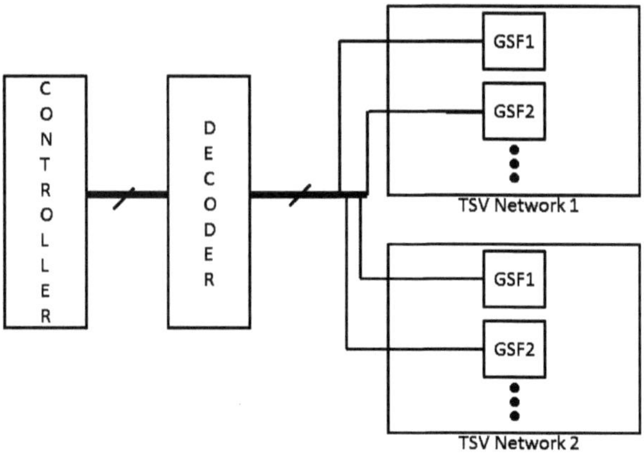

Fig. 4.3 Overview of an example control architecture

A new signal, namely the "open signal", is added; it determines whether the output Q floats or takes the value stored in the flip-flop.

In the GSF design, shown at gate-level in Fig. 4.2b and at transistor-level in Fig. 4.2c, two cross-coupled inverters are used to store data. Transmission gates are inserted between the cross-coupled inverters and at the input (D) and output (Q) of the flop itself. The widths of the transistors in the first cross-coupled inverter stage are greater than the widths of those in the second cross-coupled inverter stage such that the second stage takes the value of the first stage when the buffer between them is open and they are in contention. An internal inverter buffer is added before the output transmission gate such that the gated scan flop can drive a large capacitance on its output net without altering the value held in the flop. The "open" signal controls the final transmission gate.

It is important to distinguish between sending and receiving TSVs and their impact on test circuitry. Sending TSVs are TSVs that are being driven by logic prior to bonding, whereas receiving TSVs are floating before bonding and meant to drive logic on their associated die. In both cases, the GSF can be utilized. For the testing of sending TSVs, the GSF is used to drive the TSV during probing. In functional mode, the gate remains in its low-impedance state. In the case of receiving TSVs, the GSF also drives the TSV during test. However, in functional mode, the gate remains in its high-impedance state because the TSV will be driven by logic on another die. The functional output of the GSF associated with receiving TSVs, if needed, can be connected to node "f" as denoted in Fig. 4.2b.

A controller is needed for determining which gates are open in the TSV network at any given time. One such controller may be a centralized gate controller that is routed through a decoder to control gates in each TSV network simultaneously as shown in Fig. 4.3. Because each network is contacted by a separate probe needle, the TSVs in one network can be tested in parallel with the TSVs in each other network. Each TSV is driven by its own GSF.

For the example controller shown in Fig. 4.4, a synchronous up counter based on J/K flip-flops is used that can also be utilized as a shift register. The example

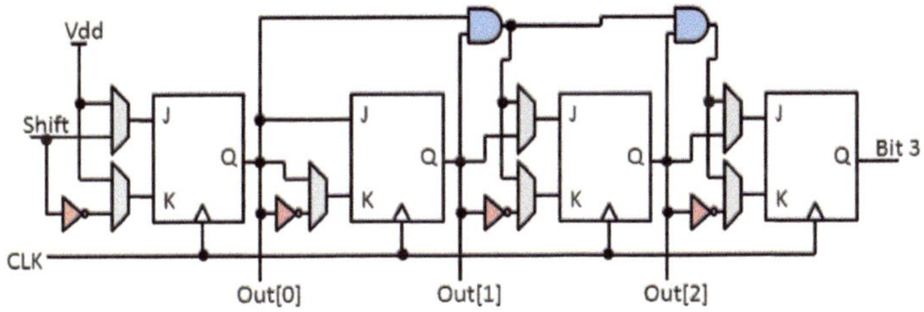

Fig. 4.4 A shift counter

Fig. 4.5 A charge sharing circuit

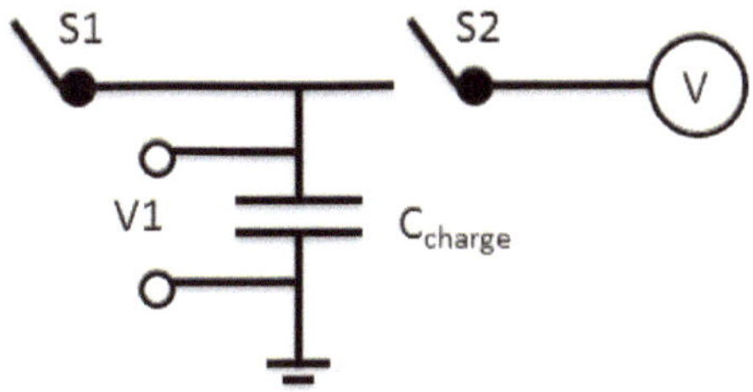

controller includes four bits; it only needs $\log_2(n)$ bits, where n is the number of gated scan-flops in the largest TSV network during test. During normal operation, the controller counts up, cycling through the gated scan-flops in each network for test. If a specific TSV must be tested or special test codes must be sent to the decoder, then the appropriate data can be shifted into the controller. A limitation of using a central controller is that outputs from the decoder must be routed to each TSV network. However, because it is only necessary to have as many wires leaving the decoder as there are TSVs in the largest network, routing can be greatly simplified, especially when compared to BIST techniques.

To determine the capacitance of the TSV network and the resistance of each TSV, the probe needle must be equipped with an active driver and a method of detection. In order to keep this circuitry simple, a design such as the one shown in Fig. 4.5 can be used. This design consists of a DC source with a voltage on the order of the circuit under test. A switch, $S2$, is used to connect or disconnect the source from a capacitor (C_{charge}) of known capacitance. The voltage across the capacitor is continuously monitored through a voltmeter. A second switch, $S1$, allows the capacitor to be connected or disconnected from the probe needle itself.

It should be noted that the above charge sharing circuit facilitates design and analysis in HSPICE. In practice, this circuit can be prone to measurement error caused by leakage currents. Therefore, an AC capacitance measurement method can be used to mitigate the effects of leakage, for example with a capacitive bridge [51].

While digital testers are usually not equipped with the drivers and sensors needed to measure capacitance, analog and mixed-signal testers are known to have these capabilities [52]. Because pre-bond TSV testing requires accurate analog measurements (digital measurements are not feasible unless more complete functionality and I/O interfaces are available), it is necessary to either add

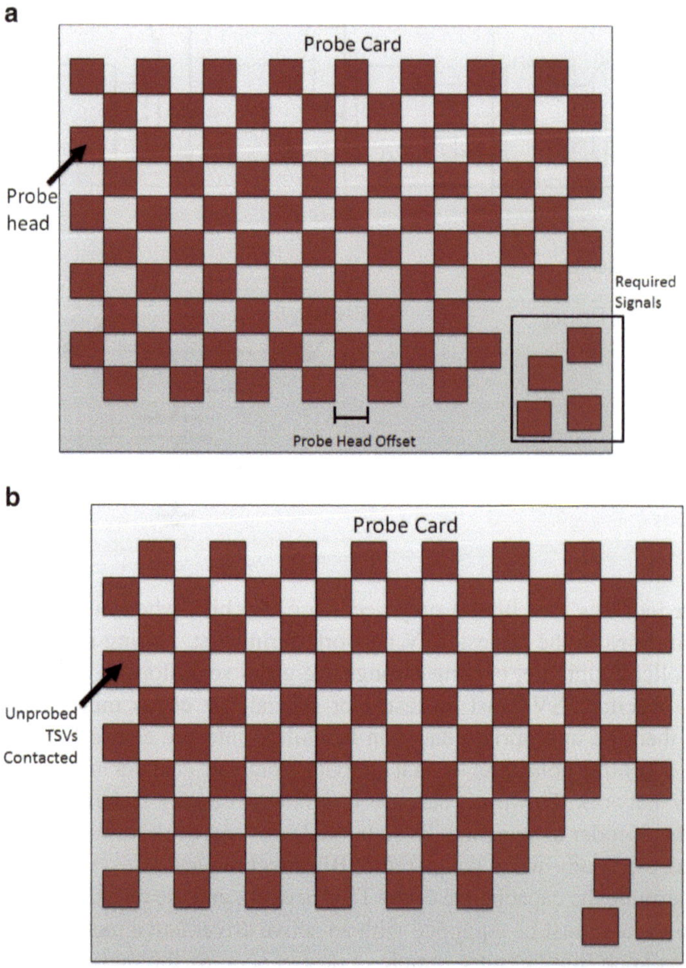

Fig. 4.6 Two configurations of a probe card for TSV testing. (**a**) Configuration 1. (**b**) Configuration 2

capacitance-sensing circuits and drivers to digital testers or to utilize analog testers for pre-bond TSV defect screening.

In order to contact all TSV matrices, the probe card has to be moved at least once. In order to reduce the number of times the probe card must be moved (and ensure a one-time-only movement), a design such as that shown in Fig. 4.6 can be utilized. By offsetting the probe needles as shown, the probe card has to be shifted up or down only once in order to contact all the TSV networks. In Configuration 1, probe needles on the probe card contact some groups of TSVs, with adjacent TSV networks lacking a probe needle to contact. Once the probe card is moved to Configuration 2, the probe needles contact the previously untested TSVs. Contacting

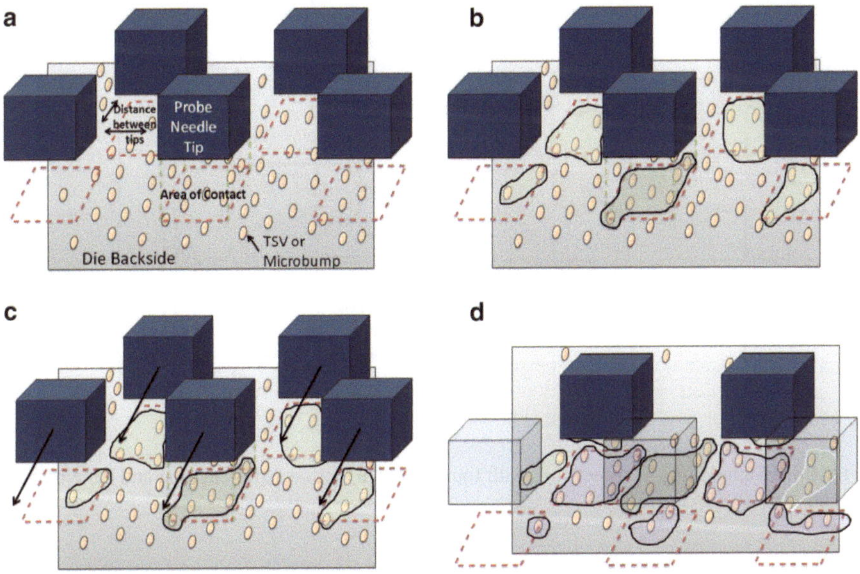

Fig. 4.7 Example to illustrate the probing of TSV networks. (**a**) First configuration. (**b**) First contact. (**c**) Shift. (**d**) Second contact

and supplying critical signals such as power and clocks to the die during test may require special probe needles to be added to the probe card; these are shown as required signals in Fig. 4.6 and are not placed in the same configuration as the rest of the probe needles. It is assumed that these will be individually contacting TSVs with large probe pads added to them.

In order to illustrate how contact is made with all TSV networks on a die, Fig. 4.7 shows a partial example of two rows of probe needles above a die with TSVs. The TSVs are spaced in an irregular manner in this example and it is assumed but not necessary that the TSVs have microbumps. Figure 4.7a shows the initial configuration of the probe card and probe needles. In Fig. 4.7b, the probe card is lowered and contact is made between the probe needles and the highlighted TSVs. Each group of TSVs contacted by one of the probe needles comprises a TSV network. The probe card is then lifted and shifted to its second configuration as in Fig. 4.7c, contacting the newly highlighted TSVs. As shown in Fig. 4.7d, a row of TSVs can be completely contacted with a single movement of the probe card. The probe card design attempts to limit the number of contacts the probe needle makes with each TSV to minimize damage that may occur during test, such as scrubbing. To prevent a single TSV from being contacted more than once, or in more than one

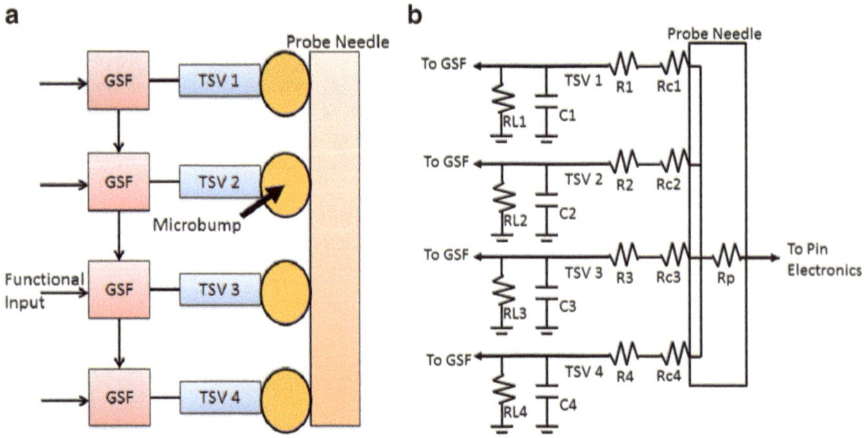

Fig. 4.8 Example of a TSV network with four TSVs: (**a**) visual representation and (**b**) electrical model

TSV network during test, additional control signals can be included in the controller to close the gates for all TSVs tested in the first test period during the second test period, and vice versa.

4.2.1 Parametric TSV Testing Via Probing TSV Networks

A TSV can be modeled as a wire with both a resistance and a capacitance. While a TSV may be manufactured from a number of different materials, copper is often used for metal layers and polysilicon may be a non-metal alternative. The resistance of a TSV made from copper with a 2–5 μm diameter and 5 μm height is 80–200 mΩ. For a polysilicon TSV with a 28–46 μm diameter and 50 μm height, the resistance is 1.3–5.0 Ω [53]. The capacitance of a copper TSV with a 1–10 μm diameter and 30–100 μm height is 10–200 fF [26].

A probe needle makes contact with a number of TSVs at a time, as seen in Fig. 4.8a. The TSVs are connected to gated scan-flops, which are connected to form a scan chain. This circuit is modeled as seen in Fig. 4.8b. The probe needle has a known resistance R_p and a contact resistance (Rc_1–Rc_4) with each TSV. The contact resistance depends on the force with which the probe needle contacts each TSV, and may differ per TSV. Each TSV has an associated resistance (R_1–R_4) and capacitance (C_1–C_4). Furthermore, a leakage path modeled by a resistance (RL_1–RL_4) exists between each TSV and the substrate. The value of interest is the net capacitance C_{net}, which is the combined capacitance of all of the TSVs in parallel. C_{net} can be expressed as:

$$C_{net} = C_1 + C2 + \cdots + C_n.$$

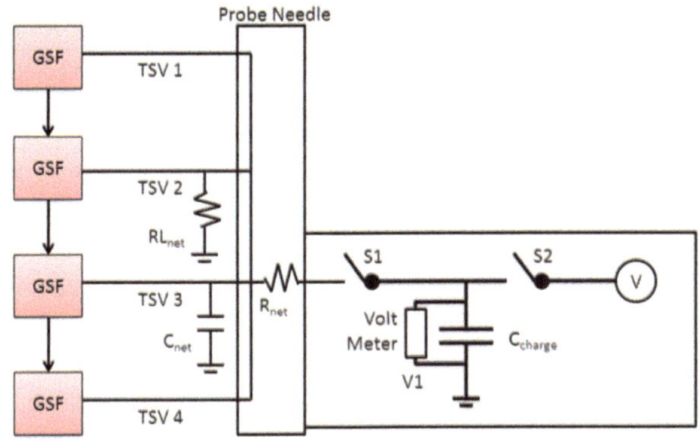

Fig. 4.9 TSV network with charge-sharing circuit

The net resistance, R_{net} is then the equivalent of the probe resistance, contact resistances, and TSV resistances, calculated as follows:

$$R_{net} = R_p + \left(\frac{1}{R_1 + R_c} + \frac{1}{R_2 + R_c} + \cdots + \frac{1}{R_n + R_c} \right)^{-1}.$$

The net leakage RL_{net} is simply all leakage resistances added in parallel.

4.2.1.1 Capacitance Measurements

The net capacitance must first be determined to characterize each TSV. From this measurement the capacitance of each TSV can be estimated and their respective resistances measured. The charge-sharing circuit of Fig. 4.5 is connected to the probe needle which shorts together multiple TSVs as shown in Fig. 4.9. There are three steps involved in measuring the net capacitance:

- Discharge the TSV network by loading a 0 into all gated scan-flops and then opening their gates. During this step, switch $S1$ is open. The charge sharing circuit is disconnected from the TSV network, and switch $S2$ is closed in order to charge capacitor C_{charge} to a known voltage V.
- Close all gated scan-flops and open switch $S2$. Close switch $S1$ to connect capacitances C_{charge} and C_{net}. This sets up the charge-sharing network as C_{charge} is discharged into C_{net}.
- Monitor the rate of change of $V1$ through the volt meter until it falls below a certain level. This level corresponds to the rate of change in a simulated charge curve that has reached 1% of its maximum charge during discharge. Once this rate is reached, then a final measurement of voltage $V1$ across capacitor C_{charge} is taken.

Once these steps are completed, the value of C_{net} can be determined from known values using the following charge-sharing equation:

$$C_{net} = C_{charge} \cdot \frac{(V - V1)}{V1} \tag{4.1}$$

From the network capacitance, the average capacitance of each TSV can be determined by dividing the network capacitance by the number of TSVs in the network. In this respect, the presence of fewer TSVs in the network will allow for a higher resolution in capacitance measurements, although this is not the case for resistance measurements or stuck-at/leakage tests (described below). Among the TSV defect types described in [38], only one of the pre-bond-testable defects results in capacitance changes (as opposed to resistance changes). This is the pinhole defect, which may also be detected through leakage tests. Although capacitance measurement using this method yields only an average value, significant increases in capacitance can be readily detectable if the number of TSVs in a network is not too large.

4.2.1.2 Resistance Measurements

The bulk of TSV defects that can be tested pre-bond result in increased TSV resistance. For this reason, it is important that a pre-bond test be capable of accurately measuring TSV resistance. In order to measure resistance, the charge-sharing circuit of Fig. 4.5 is once again utilized. The capacitor C_{charge} will be charged through each TSV, and the time needed to charge the capacitor to a chosen voltage (for example, 99 % of V_{dd}) is recorded. Long charge times increase the resolution in resistance measurement, but they lead to higher test time. As a tradeoff, smaller voltage levels (such as 90 % of V_{dd}) can be used to reduce test times if the resolution is acceptable—see Table 4.1 in Sect. 4.2.2 for more information.

The above measurement can be carried out by recording the start time when the control signal is asserted for the TSV under test to open, then the end time can be measured when $V1$ reaches the desired voltage. In order for resistance to be measured, the probing device must first be calibrated using a non-faulty TSV in a TSV network. This calibration can be done off-chip prior to testing any circuits, for example by using a dummy silicon chip with double-ended TSVs in which the TSVs themselves can be completely characterized. One or more TSVs on this calibration chip can be used to calibrate the equipment. The charge time of C_{charge} in this environment is determined, and charging times on the test floor are then viewed in relation to the calibrated time.

The test begins by loading all of the gated scan-flops with 1 and discharging the TSV network using the probe. Switch $S2$ is opened and switch $S1$ is closed such that the capacitor C_{charge} is discharged as well. One of the gated scan-flops is then opened, allowing the scan-flop to charge C_{charge} through its connected TSV. When $V1$ reaches the pre-determined voltage, the time to charge C_{charge} is recorded. It is then compared to a calibrated charge curve for a non-faulty TSV. This process of charging and discharging continues for each TSV, which can be completed quickly by incrementing the controlling counter to open each subsequent TSV.

4.2.1.3 Leakage Tests

Leakage tests are an averaged measurement per TSV, similar to capacitance tests described earlier. In order to perform a leakage test, C_{charge} is disconnected from the TSV network and the network capacitance C_{net} is charged through the GSFs to V_{dd}. Next, all gates are switched to their high impedance state and the TSV network is left floating for a chosen period of time. After this period, a voltage measurement of the network is taken through the probe, and the change in voltage over the floating time period is determined. This is then compared to a calibrated curve to determine the leakage of the network.

4.2.1.4 Stuck-At Tests

Stuck-at and leakage tests, in which leakage is high enough to be similar to stuck-at faults, can be performed together and in parallel under this scheme. For strong stuck-at 0 faults or leakage with low resistances to ground, the TSV network can be charged with the gated scan-flops closed and its voltage measured. If the rate of discharge is abnormally high, it can be inferred that a stuck-at 0 fault or a leakage-inducing defect exists on at least one of the TSVs. A parallel stuck-at 1 test can be performed by discharging the TSV network with the gated scan-flops closed and measuring the voltage on the net.

Individual stuck-at tests can also be performed quickly. This is done by loading the scan-chain with a pattern of alternating ones and zeros. The value on the gated scan-flop on the first flop in the control sequence for the TSV network determines whether the net is first charged or discharged. Then, each GSF is opened in sequence, making alternating assertions of high or low. The pattern is then shifted by one and the process is repeated.

4.2.2 Simulation Results for Pre-bond Probing

Experimental results are presented for a TSV network of 20 TSVs modeled in HSPICE. The number 20 was determined based on the relative diameter and pitch of probe leads and TSVs. Unless otherwise stated, the resistance of each TSV and contact resistance is 1 Ω and the TSV's associated capacitance is 20 fF. These numbers are based on data reported in the literature [26, 53]. The probe needle resistance is set at 10 Ω. This value is several Ω higher than contact resistances seen with probe cards today [54,55] to account for the low contact force needed and unusually small features probed in our scheme. TSV leakage resistances for non-faulty TSVs were 1.2 TΩ, corresponding to a leakage of 1 pA at a 1.2 V V_{dd} [83]. The transistors were modeled using predictive low-power 45 nm models [56]. Transmission-gate transistor widths were set to 540 nm for PMOS and 360 nm for NMOS. These larger widths were chosen such that the gate, when open, would have

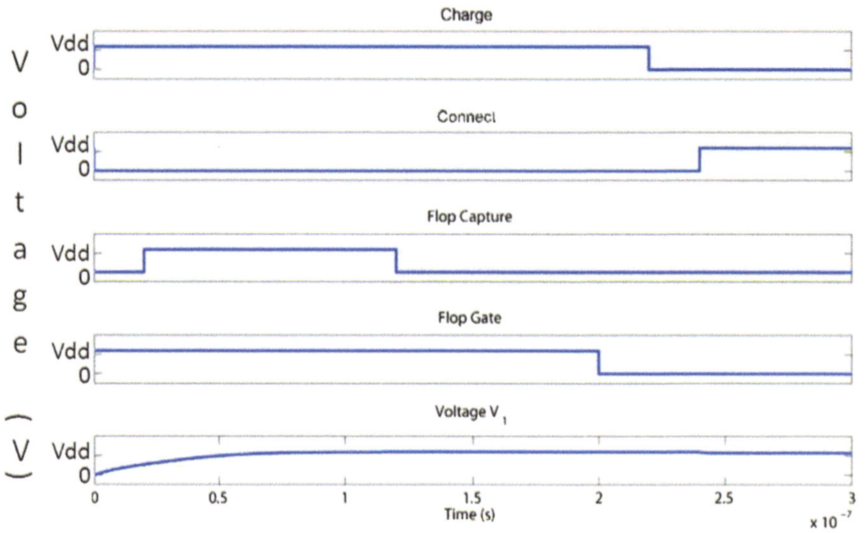

Fig. 4.10 The process of net capacitance measurement

little impact on signal strength. A strong and a weak inverter were used, with the strong inverter having widths of 270 nm for PMOS and 180 nm for NMOS, and the weak inverter having 135 nm for PMOS and 90 nm for NMOS. These were chosen such that the majority of transistor W/L ratios were 2/1 for NMOS and 3/1 for PMOS. The charge-sharing capacitor C_{charge} was modeled at 10 pF, chosen to be an order of magnitude larger than the fault-free capacitance of the TSV network. This is sufficiently large to achieve good resolution in measurement without being so large that charge times are unreasonable or that leakage becomes a significant issue. The power supply voltage V_{dd} for both the probe electronics and the circuit under test was set at 1.2 V.

Inductance is not included in the model for two reasons. First, modern probe cards have little parasitic inductance on the probe tips [57]. Second, sampling takes place in the pin electronics and not through the TSV network itself, so pin electronics are the limiting factor for high-speed sampling and not the TSV network or its contact with the probe needle. Probes capable of GHz sampling frequencies have been available for some time [58].

Figure 4.10 demonstrates the process of net capacitance measurement, with high signals meaning that a gate is open or a switch is closed. To begin, switch S2 is closed, charging C_{charge} to V. During this time, the gated scan-flops capture a 1 (the Flop Capture signal captures on the falling edge). The flop gates are open as denoted by the Flop Gate signal. The flop gates are then closed, S2 is opened, and switch S1 is closed to begin charge-sharing. C_{charge} then begins discharging, and the voltage is measured after 250 ns when it has settled to 1.15 V. Using Eq. 4.1 and the subsequent division step, each TSV capacitance can be determined to be 20.25 fF, very close to the actual value of 20 fF.

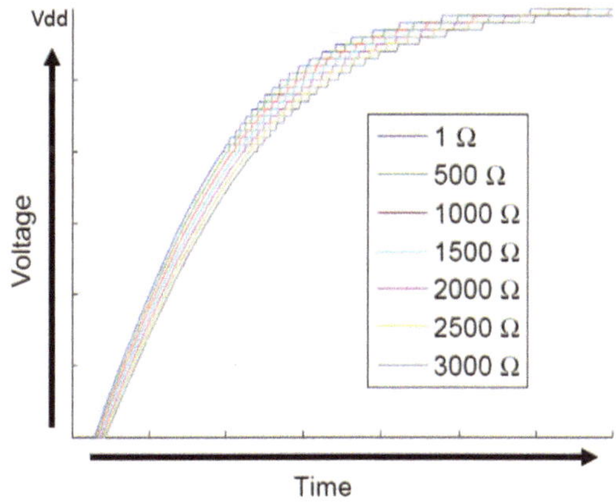

Fig. 4.11 Capacitor charging through TSVs of varying resistance

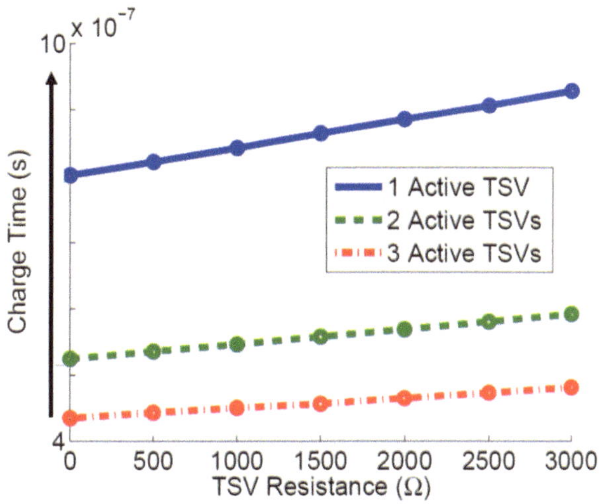

Fig. 4.12 Capacitor charge time through one, two, and three TSVs to 0.99 V_{dd}

The determination of C_{net} is a robust measurement. Because the charge sharing system set up between C_{charge} and C_{net} is allowed to settle before measurements are taken, TSV and contact resistance do not affect the result. Only relatively high leakage currents prevent measurements, because the change in voltage of C_{charge} will remain large until the capacitor is discharged. For example, a 100-point Monte Carlo simulation was conducted assuming a Gaussian process variation with a 3-σ value of 20 % around nominal TSV and leakage resistance values. TSV capacitances ranged from 10 to 50 fF, with a total net capacitance of 550 fF. In every simulation, C_{net} was calculated to be 569 fF regardless of process variation.

Figure 4.11 shows the charging behavior of capacitor C_{charge} through one TSV in the TSV network. The TSV resistance varies from 1 to 3,000 Ω in 500 Ω intervals. V_1 is recorded when the voltage across C_{charge} has reached 99 % of V_{dd}, or 1.19 V. As shown, small variations in resistance alter the charge curve in measurable ways. Figure 4.12 shows the charge time to reach this voltage level for each TSV resistance

Table 4.1 Resolution of TSV resistance measurements with a 500 MHz and 1 GHz sample rate for a fault-free 1 Ω and faulty 500 Ω TSV at various voltage levels

Chosen voltage level	Minimum detectable resistance change	
(percentage of V_{dd})	At 1 GHz (Ω)	At 500 MHz (Ω)
99	24.3	48.6
95	40.4	80.8
90	55.6	111.2
60	161.3	322.6
50	221.2	442.4
40	324.7	649.4
10	2777.8	5555.6

and for one, two, or three TSVs under test in parallel. As can be seen, there is a linear relationship between capacitor charge time and the resistance of the TSV under test. For the capacitance value of 10 pF, each 500 Ω increment increase in TSV resistance results in about a 20 ns increase in charge time when considering only one TSV under test. Assuming a sample rate of 1 GHz and calibration at 1 Ω (the first waveform), a resolution r of about 25 Ω is achieved. In other words, each increase in charge time of 1 ns above the calibrated charge time corresponds to a 25 Ω increase in resistance on the TSV under test. In this scheme, smaller values of r preferable. Higher resolutions can be achieved at the cost of longer charge times by increasing the capacitance of C_{charge}. However, if the capacitance is too large then leakage could become a significant source of error. Generally, the resolution of measurement can be determined using the formula $\frac{\Delta\Omega}{S \cdot \Delta T}$ where ΔT is a change in charge time, $\Delta\Omega$ is the TSV resistance difference for that charge time, and S is the sample rate.

Table 4.1 shows the resolution of TSV resistance measurements at 500 MHz and 1 GHz sample rates for different chosen voltage levels, assuming a fault-free TSV with 1 Ω resistance and a faulty TSV with 500 Ω resistance. For example, the resolution achieved when charging C_{charge} to 99 % of V_{dd} implies that resistances as small as 24.3 Ω above the nominal 1 Ω fault-free resistance can be detected. The lower the entries in the second column of Table 4.1, the larger the resolution and detectability of TSV defects. It can be seen that as the voltage level to which C_{charge} is charged decreases, the resolution achievable by resistance measurements also decreases.

Table 4.2 shows the calculated TSV resistance values of several faulty TSVs using the calibration curve for a single TSV of Fig. 4.11. As can be seen, high accuracy is achieved for a range of faulty resistances. Higher resolutions are achieved in the 400–600 Ω range, although this is based on a curve calibrated only at every 500 Ω. It is expected that more data points in the calibration curve would lead to more accurate results at other resistance values.

The test time for TSV resistance measurements can be estimated from the voltage level to which C_{charge} is charged and the number of TSVs and TSV networks that must be tested. For example, consider a die with 10,000 and 20 TSVs per

Table 4.2 Measurement accuracy at varying faulty TSV resistances

Actual resistance (Ω)	Measured resistance (Ω)	Percent difference
100	110.8	10.8
200	207.3	3.7
300	304.3	1.4
400	401.8	0.5
500	499.1	0.2
600	596.8	0.5
700	695.0	0.7
800	793.4	0.8
900	891.8	0.9
1000	990.8	0.9

network, for which C_{charge} is charged to 99 % of V_{dd}. Due to bandwidth and current limitations of the probe card, it is assumed that only 100 TSV networks can be tested in parallel at a time. From simulations of fault-free TSVs, maximum currents of 46 μA are sunk through each probe needle during resistance measurement. This is well within the current limits of the smallest probe needles [59] (≈120 mA for tip diameter of 1.0 mil and ≈400 mA for tip diameter of 5 mil), and it is thus likely that in these circumstances, more than 100 TSV networks could be tested at a time. The time required for measuring the resistance of all TSVs in this example is 80 μs, not including the time required to move the probe card.

It is also possible to test the TSV resistance of multiple TSVs in parallel at the cost of resolution of the result. Figure 4.12 shows charge times when two or three parallel TSVs are under test. In each case, the resistance for all TSVs in the group tested in parallel increases from 1 to 3,000 Ω in increments of 500 Ω. A loss of resolution is experienced on two fronts. The first is that the difference in charge times between the chosen TSV resistances decreases to 10 ns for two TSVs in parallel and 5 ns for three. This loss of resolution can be overcome to an extent with a larger capacitance C_{charge}, although larger capacitances are more susceptible to error from leakage. The second reason for loss of resolution lies in the averaging that must take place between the resistances of the TSVs tested in parallel. This problem cannot be easily alleviated using the proposed method. Some averaging may be desirable in the test environment, in which case it would be faster to test groups of TSVs in each TSV network in parallel with one another. The controller can be designed appropriately.

The robustness of TSV resistance measurements in a 20-TSV network under process variations is examined next. The TSV under test is considered to have a resistive fault with a total resistance of 50 Ω. Resistances on the other TSVs in the network are simulated with a Gaussian distribution in which 3-σ is a 20 % spread from the nominal value of 1 Ω. All TSV capacitances are simulated with a similar Gaussian distribution using a nominal value of 20 fF, and leakage resistances are distributed around a nominal 1.2 TΩ. Charge times are then compared to a calibrated curve. As can be seen from a 100-trial Monte Carlo simulation in Fig. 4.13, the

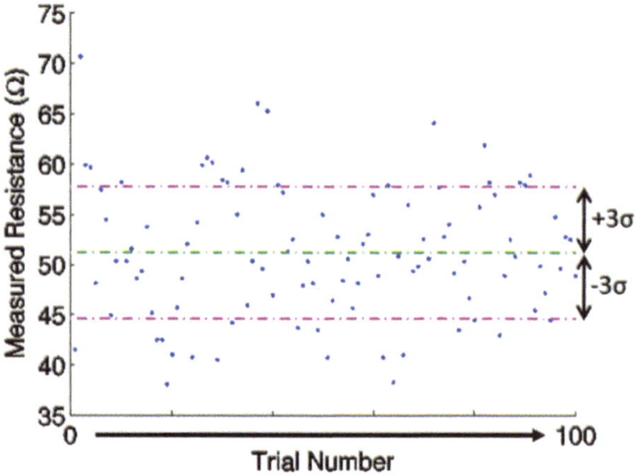

Fig. 4.13 Hundred-point Monte Carlo simulation of TSV resistance measurements with 20 % variation in the TSV resistance, leakage resistance, and capacitance of fault-free TSVs

resolution of resistance measurements remains high under process variations, with a mean measurement of 51.2 Ω and a standard deviation of 6.6 Ω.

The accuracy of TSV resistance measurements in a TSV network where more than one TSV is faulty is explored. The Monte Carlo simulations of Fig. 4.13 are repeated, this time assuming that each TSV is defective with a Gaussian probability density function. For this example, let the 3-σ value for defective TSV resistance be 100 Ω around a 150 Ω nominal value. It is assumed that the 3-σ value of the TSV capacitance under process variation is 30f fF, with a nominal value of 20 fF. The leakage resistance 3-σ value was 400 GΩ around a 1.2 TΩ nominal value. Figure 4.14 presents results for a 100-trial Monte Carlo simulation in this scenario. Good resolution in resistance measurements continues to be achieved, with a mean of 141 Ω and a standard deviation of 54 Ω. While the defective TSVs are severe, their impact on resistance measurements is reduced because a capacitance C1 that is orders of magnitude larger than the TSV capacitance is chosen. The charge time of this capacitor dominates the much smaller changes in charge time caused by varying RC values in the TSV network.

Finally, the accuracy of TSV resistance measurements in a TSV network when the contact resistance varies between TSVs is examined. Many probe needles are not flat, for example they may terminate at tapered plateaus. Three different models of TSV contact resistance are explored. The first (static) profile assumes that contact resistance is independent of TSV location in the TSV network, with a Gaussian distribution for contact resistance with a 3-σ value of 10 Ω around a 40 Ω expected value. The second (linear) profile increases contact resistance linearly within the TSV network the further a TSV is from the center of the theoretical probe needle. The linear profile varies contact resistance with a Gaussian function per TSV with a 3-σ value of 2 Ω around a 5 Ω expected value for the innermost TSVs to a 3-σ value of 15 Ω around a 30 Ω expected value for the outermost TSVs. The last (exponential) profile increases contact resistance exponentially within the TSV network the further a TSV is from the center of the network. The exponential profile

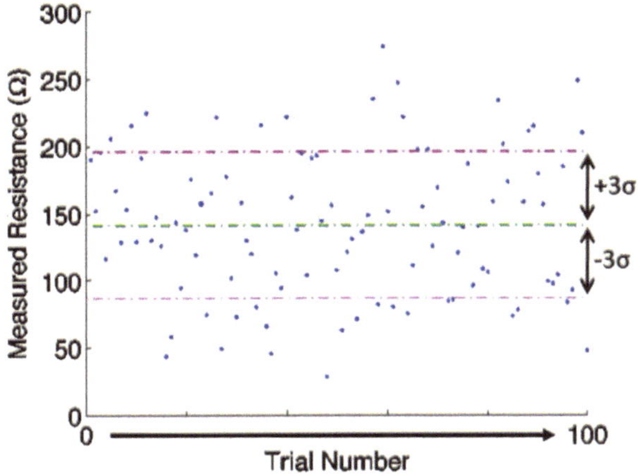

Fig. 4.14 Hundred-point Monte Carlo simulation of TSV resistance measurements for multiple TSV resistive, leakage, and capacitive defects and under process variations

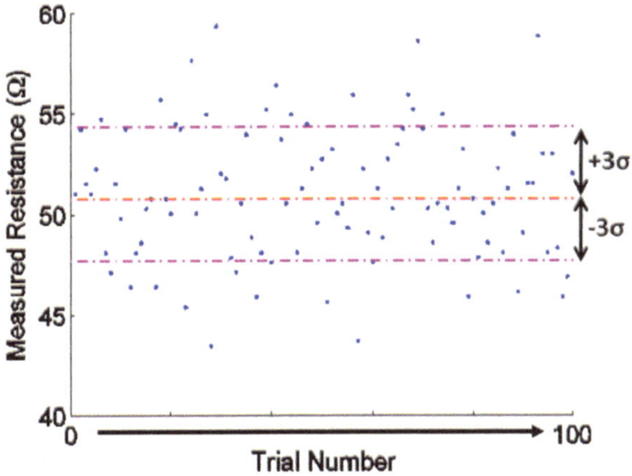

Fig. 4.15 Hundred-point Monte Carlo simulation of TSV resistance measurements for a static profile of TSV contact resistances

varies contact resistance with a Gaussian function per TSV with a 3-σ value of 5 Ω around a 5 Ω expected value for the innermost TSVs to a 3-σ value of 20 Ω around a 100 Ω expected value for the outermost TSVs. A 100-point Monte Carlo simulation was performed for each profile in an attempt to measure a 50 Ω faulty TSV. Contact resistance is additive to TSV resistance, so the expected value of the contact resistance is subtracted from the measurement to obtain the TSV resistance. The results for the static profile are shown in Fig. 4.15, with a mean measured faulty resistance value of 50.8 Ω and standard deviation of 3.3 Ω. The simulation results for a linear profile are shown in Fig. 4.16, and results for an exponential profile are shown in Fig. 4.17. The faulty resistance measurements for these simulations were a mean of 50.9 Ω and standard deviation of 0.7 Ω for the linear model and a mean of 50.8 Ω and standard deviation of 1.7 Ω for the exponential model. As can be inferred, as long as the expected value of contact resistance is near to the actual contact resistance, accurate measurements of TSV resistance can be obtained due to the additive nature of contact resistance.

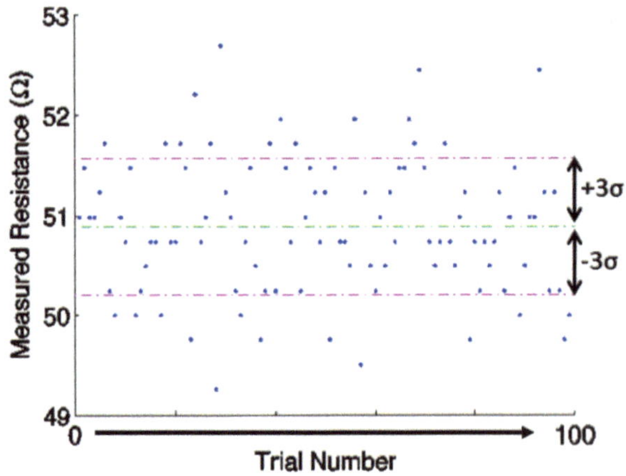

Fig. 4.16 Hundred-point Monte Carlo simulation of TSV resistance measurements for a linear profile of TSV contact resistances

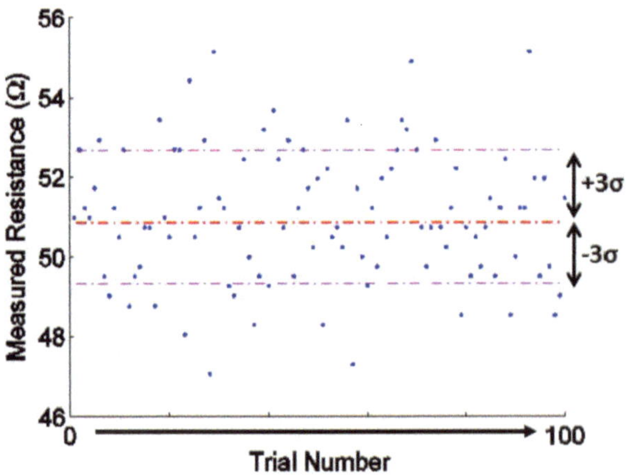

Fig. 4.17 Hundred-point Monte Carlo simulation of TSV resistance measurements for an exponential profile of TSV contact resistances

Similar to the calibration curves for TSV resistance measurements, calibration curves can be determined for leakage resistance as shown in Fig. 4.18. This calibration plots the voltage of C_{net} after 8 μs of the TSV network left in a floating state on the x-axis. The y-axis consists of the corresponding total leakage resistance RL_{net}. Due to the non-linear property of capacitance discharge, a calibration curve was created from this data using a logarithmic fit in base 10.

The effect of process variation on leakage resistance measurements is shown in Fig. 4.19. As before, 100-point Monte Carlo simulations were performed. A Gaussian distribution was used with TSV resistance, leakage resistance, and TSV capacitance varying with a 3-σ of 20 % around their nominal values. One faulty TSV was assigned a leakage resistance of 100 MΩ. As can be seen, this leakage was accurately determined in network resistance measurements, with a mean of 100.5 MΩ and a standard deviation of 1.4 MΩ.

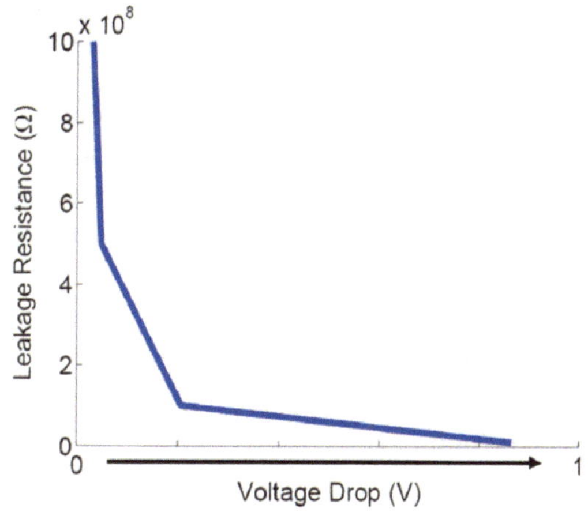

Fig. 4.18 Plot of RL_{net} versus the voltage change of C_{net} over 8 us

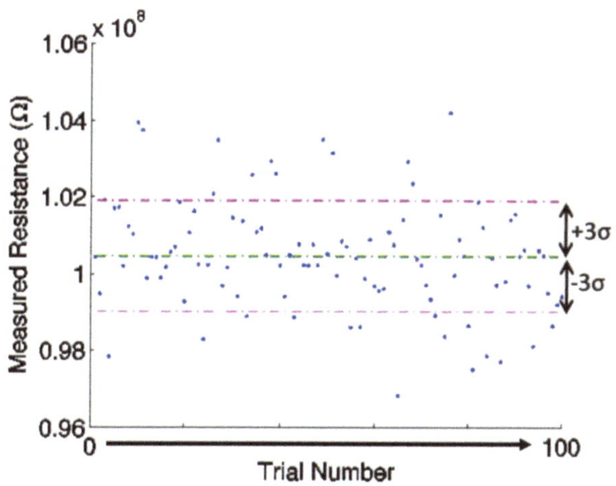

Fig. 4.19 Hundred-point Monte Carlo simulation of leakage resistance measurements with 20 % variation in the TSV resistance, leakage resistance, and capacitance of fault-free TSVs

4.2.3 Limitations of Pre-bond TSV probing

This section presented DFT and ATE-compatible measurement methods for pre-bond probing of TSVs. Several enhancements to this basic approach are possible. The need for averaging across all TSVs in a network implies that the resolution in capacitance measurement may be reduced in larger networks. This problem is not severe, however, as resistance and leakage tests can be used to detect most pre-bond defects in TSVs presented in [38]. The proposed method also requires that more than one contact be made with the thinned wafer during testing; therefore, it is important to minimize the number of times the probe needle must be moved during testing to avoid damage and to reduce test time. Pinpointing which TSVs in a network contribute to averaged errors, such as capacitance and leakage, is difficult with this method. Therefore, it is an open problem to identify and repair faulty TSVs.

The addition of an analog tester to the test flow and the need for new probe card designs add to the test cost. Moreover, the need to move a die between testers will inevitably increase test cost. Next-generation testers might provide the required measurement capabilities. Furthermore, there is an area overhead associated with the proposed architecture, although existing test structures are reused where possible. Ultimately, the yield of TSVs versus the cost of test will have to be considered before test architecture decisions are made.

4.2.4 Summary

Section Summary

- Die-level boundary registers can be replaced with gated scan-flops to enable parametric TSV test when contacting multiple TSVs simultaneously through a single probe needle.
- Parametric TSV test can be performed to determine TSV capacitance, leakage, and resistance.
- TSV capacitances and leakages must be averaged between TSVs in a TSV network, resulting in a loss of test resolution for large networks.
- Pre-bond TSV testing yields reliable and accurate results even in the presence of process variations and multiple defective TSVs and regardless of defect location on the TSV.
- On-die area overhead can be low, because test circuits and analog components are moved off-die to pin electronics.
- The need for probe equipment increases the cost of test and can require longer test times when compared to BIST techniques.

4.3 Reducing Test Time Through Parallel TSV Test and Fault Localization

Section 4.2 introduced a DFT architecture and techniques for pre-bond probing of TSVs for thinned wafers. The key idea in this probing technique is to use each probe needle to simultaneously contact multiple TSVs, forming a "TSV network". Recall Fig. 4.12, which demonstrated that, within a TSV network, multiple TSVs can be tested in parallel to decrease test time, but at the expense of reduced resolution of analog measurements for each TSV. In order to carry out diagnosis and TSV repair using methods such as those described in [40, 84], it is necessary to identify the individual TSVs that are faulty out of those TSVs that are tested simultaneously within a TSV network.

In this section, an algorithm will be developed for designing parallel TSV test sessions such that TSV test time is reduced and a given number of faulty TSVs

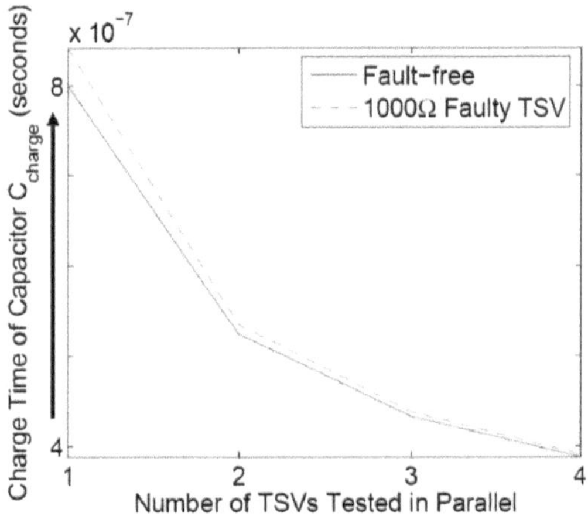

Fig. 4.20 Reduction in capacitor charge time when driven through multiple TSVs

within the TSV network can be uniquely identified under parallel test. The algorithm returns the sets of TSVs to test in parallel within a network. The algorithm is efficient and fast, and therefore can be used as a subroutine in a more general algorithm for optimizing TSV networks.

As the number of TSVs tested in parallel increases, there is a reduction in charge time for C_{charge} for both fault-free networks and networks with a single 1,000 Ω faulty TSV, as shown in Fig. 4.20. The difference in charge times between a faulty and fault-free network decreases as more TSVs are tested in parallel (Fig. 4.12), which adversely affects resolution. Therefore, while larger TSV networks allow more TSVs to be tested simultaneously to reduce test time, the number of TSVs per test session cannot be increased beyond a limit due to resolution constraints.

Consider a TSV network consisting of six TSVs. A simple solution for testing is to test each TSV individually, resulting in six test sessions. However, significant savings in test time can be achieved if multiple TSVs are tested in parallel and the repair mechanism can target the faulty TSVs in an individual network. Consider an upper limit m on the number of faulty TSVs that need to be identified in each network. This limit would be defined based on the capability of the on-die repair architecture or on the desired level of fault localization. If the goal of TSV test is to pinpoint only one faulty TSV in a TSV network, then each TSV i needs to be in two test groups such that the TSVs in the first network for i are different from the TSVs in the second network for i. In other words, let S_1 and S_2 be the set of TSVs that are grouped with i in the first and second test sessions, respectively. Then $S_1 \cap S_2 = \emptyset$. If $m = 2$, then three unique test sessions are needed for each TSV to distinguish m faulty TSVs. Thus, the number of unique test sessions needed per TSV is equal to $m + 1$.

The above reasoning can be explained conceptually using the example of $m = 2$. Any fault-free TSV i may be in a test session with faulty TSV f_1, faulty TSV f_2, or in a test session without either f_1 or f_2. Hence, i can be in three different test sessions. In a worst-case scenario, two of the test sessions for i will contain one of

Table 4.3 Parallel tests needed for pinpointing defects for one and two faulty TSVs in a 6-TSV network

Test session number	Test groups created ($m = 1$)	Test groups created ($m = 2$)
1	{1,2,3}	{1,2,3}
2	{1,4,5}	{1,4,5}
3	{2,4,6}	{2,4,6}
4	{3,5,6}	{3,5,6}
5	–	{1,6}
6	–	{2,5}
7	–	{3,4}

f_1 and f_2. Thus, these two sessions will fail. However, the third test session can then contain neither f_1 nor f_2 and will pass, indicating that i is fault-free. It is important to note that the condition presented above is sufficient but not necessary, as it is possible that none of the test sessions for i contain f_1 or f_2.

Table 4.3 shows an example of test groups that can be designed for a network of six TSVs in which at most three TSVs can be tested in parallel. Column 2 shows test groups if only one faulty TSV needs to be pinpointed, and Column 2 shows results for two faulty TSVs. When $m = 1$, the number of tests needed can be reduced by 2 with a resulting significant decrease in capacitor charging time, creating a 63.93 % reduction in test time, based on Fig. 4.20. For $m = 2$, one more test session is needed per TSV but a 31.19 % reduction in test time is still possible. If three or more faulty TSVs need to be identified per network, then for this example separate testing of the TSVs in a network is the best choice.

With the above example as motivation for parallel testing of TSVs, a formal problem statement can be developed. The problem of parallel-test-group creation for TSV networks is defined as follows.

Given the number of TSVs to be tested (T), the tester bandwidth B (the number of probe needles that can be active at one time, which determines how many TSV networks can be tested in each test period), a set P of the test times associated with testing different numbers of TSVs in parallel, the number of faulty TSVs m that must be identified per TSV network, and a minimum resistance resolution r, **determine** the parallel test sets for each TSV network in order to minimize overall test time while keeping the resolution of measurements at or above r and ensuring that up to m faulty TSVs in any given TSV network are uniquely identifiable.

4.3.1 Development of an Algorithm for Parallel TSV Test Set Design

Before describing the optimization algorithm, several constraints can be derived from the problem definition. These are done during algorithm initialization. First, for the purpose of this section, the TSVs are evenly distributed to networks, with the largest network having $\lceil \frac{T}{2B} \rceil$ TSVs. The "2" in the denominator results from the two

Fig. 4.21 The setMatrix and initial steps for a TSV network with $T = 4$ and $m = 1$

$$\begin{bmatrix} 1 & 2 & 3 & 4 \\ 2 & 3 & 4 & 1 \\ 3 & 4 & 1 & 2 \\ 4 & 1 & 2 & 3 \end{bmatrix} setMatrix$$

$$\begin{matrix} \begin{bmatrix} 1 & 1 & 0 & 0 \end{bmatrix} & \begin{bmatrix} 2 & 1 & 1 & 0 \end{bmatrix} \\ \begin{bmatrix} 1 & 2 & 3 & 4 \\ - & 3 & 4 & 1 \\ 3 & 4 & 1 & 2 \\ 4 & - & 2 & 3 \end{bmatrix} & \begin{bmatrix} - & 2 & 3 & 4 \\ - & 3 & 4 & - \\ - & 4 & - & 2 \\ - & - & 2 & 3 \end{bmatrix} \end{matrix}$$

separate test periods for all the TSV networks. This does not alter the generality of the algorithm, as the algorithm can be utilized for TSV networks of any size given the design and test constraints of an actual 3D design. The constant $numTests$ takes the value of $m + 1$, the number of test groups that are needed for each TSV. The variable $curRes$, which keeps track of the maximum number of TSVs that the algorithm tries to test in parallel with each other, is initialized to $\lceil \frac{T}{numTests} \rceil$ or r, whichever is lower. In most networks, $curRes$ will equal r, except in small networks where r is a significant fraction of the TSVs in the network when compared to the number of test groups needed for each TSV. In these cases, combining TSVs into test groups at the maximum resolution will result in sub-optimal test groups.

To keep track of each TSV and those TSVs that it has already been tested with, a $T\text{x}T$ matrix, $setMatrix$, is initialized. In this matrix, the TSVs are labeled from 1 to T, with each column offset by a value of one from the previous column, as shown in Fig. 4.21 at the top for a TSV network with $T = 4$.

A vector, $used$, is also initialized with length T, to track how many times each TSV has been used in a test group. All values in this vector are initialized to 0. A number of functions that act on $setMatrix$ and $used$ are further defined. The function **longIntersect(a)** takes a set a of columns from $setMatrix$ and returns the set of values formed from the intersection of all of the columns. This function, if acting on N columns with T TSVs, has a worst-case time complexity of $O(N \cdot T)$. The function **bestIntersection(b)** takes one column, b, of $setMatrix$ and determines which other column results in a set with the most intersections and contains the first value in each column. It will stop immediately if the number of intersections is equal to or greater than r. The complexity of this step is $O(T^2)$.

The function **updateUsed(c)** takes a set of TSV numbers, nullifies corresponding values in $setMatrix$ columns represented in the set, and increments the $used$ value for each TSV. If the $used$ value for a TSV equals $numTests$, then that column in $setMatrix$ is removed completely from consideration in future test groups. This step removes the TSV number associated with the column from all other columns. For example, Fig. 4.21 shows two different iterations of the vector $used$ (above each matrix) and $setMatrix$ for a network with $T = 4$ and $m = 1$. On the bottom left, TSVs 1 and 2 have been added to a test group together. Therefore, the $used$ value for each has been incremented to 1 and the TSV numbers have been removed from

Algorithm 1 createTestGroups(T,B,P,m,r)

```
Create and initialize setMatrix, used, numTests, curRes;
testGroups = { };
for i = 1 to T do
   while used[i] < numTests do
      inter ← bestIntersection(i);
      if (curRes ≥ 4) AND (size(inter) geq 4) then
         for each set b of curRes TSVs in inter do
            bestInter = { };
            if size(longInterb) > size(bestInter) then
               bestInter ← longInter(b);
               if size(bestInter) ≥ curRes then
                  break;
               end if
            end if
         end for
      else
         bestInter ← inter;
      end if
      if (T ∃ bestInter) AND (used(T) < numTests - 1) AND (notNull(setMatrix) < curRes - 1)
      then
         curRes ← ⌈ curRes / 2 ⌉;
         next;
      end if
      reduce(bestInter);
      testGroups ← testGroups + bestInter;
      updateUsed(bestInter);
   end while
end for
```

column 1 and 2. On the bottom right, TSV 1 is subsequently added to a test group with TSV 3. This increments $used$ for TSV 1 to 2 and for TSV 3 to 1. Because TSV 1 has reached the value of $numTests$, the entire column associated with TSV 1 is nullified and is removed from all columns. TSV 3 would also be removed from column 1; however, the column was deleted when TSV 1 was placed in its second session.

Two other functions need to be defined. The function **notNull**($setMatrix$) returns the number of columns in $setMatrix$ that have not been nullified. The function **reduce**(d) takes a vector d and reduces the number of values in it to be equal to $curRes$. This is done with respect to intersection sets. The function preserves the TSV numbers in the set corresponding to the TSVs tested and returned by **bestIntersection**().

Now, the algorithm **createTestGroups** can be described, (Algorithm 1). The algorithm begins with initializations, including the creation of $testGroups$, a set containing sets of TSVs that are tested in parallel. The algorithm runs iteratively through each TSV starting with TSV 1, assigning them to test groups until their corresponding $used$ value is above $numTests$. To determine which TSVs have not yet been tested with each other, intersections between $setMatrix$ columns are determined.

The final **if** statement in the algorithm exists to reduce the value of $curRes$ that the algorithm is trying to match to avoid sub-optimal group assignments for the final TSVs in a network. For example, consider that for a network with $T = 20$, TSVs 17, 18, 19, and 20 remain to be tested, $curRes$ is 4, and m is 1. The algorithm attempts to place all the TSVs into a test group $\{17, 18, 19, 20\}$, incrementing their $used$ values to 1. However, each $used$ value must equal 2 because $numTests$ is 2. Therefore, each TSV must then be tested individually, even after testing them together. To avoid this, $curRes$ is instead decremented to 2 and the algorithm tries again. Decrementing could continue if needed, but at 2 this yields the test groups $\{17, 18\}$, $\{17, 19\}$, $\{18, 20\}$, and $\{19, 20\}$, which reduces test time compared to testing each TSV individually.

The above procedure guarantees that $m + 1$ unique test groups are created for each TSV and that each faulty TSV can be uniquely identified. The $used$ vector ensures that each TSV is placed in $m + 1$ unique test sessions. The one exception to this rule is when a TSV is placed in a test session with no other TSVs, in which case this single test session is sufficient for determining whether or not the TSV is faulty. To ensure that each test session contains a unique combination of TSVs, the $setMatrix$ and associated column intersections identify those TSVs that have and have not been tested together. Each intersection returns those TSVs that can still be combined to form unique test sessions.

An example underlying the iterative nature of this algorithm is shown in Fig. 4.22 for $T = 6$, $m = 1$, and $r = 4$, i.e., up to four TSVs may be tested in parallel. The $used$ vector is shown on top, with $setMatrix$ below. After initialization, the value of $curRes$ is four. The first iteration adds the set $\{1, 2, 3, 4\}$ to $testGroups$, with $used$ and $setMatrix$ updated appropriately. The second set produced is $\{1, 5, 6\}$, followed by $\{2, 5\}$, $\{3, 6\}$, and finally $\{4\}$. This results in five test groups compared to six for a serial test case, and results in a test time reduction of 44.40 %.

4.3.2 *Evaluation of the* **createTestGroups** *Algorithm*

In this section, experimental results for TSV networks with varying values of T, m, and r are presented. In order to determine test times, simulations using HSPICE were done on a TSV network of 20 TSVs. The resistance of each TSV and contact resistance is 1 Ω and the TSV's associated capacitance is 20 fF. These numbers are based on data reported in the literature [26, 53]. The probe head resistance is 10 Ω. This value is several Ω higher than contact resistances seen with probe cards today [54, 55] to account for the low contact force needed and unusually small features probed in our scheme. The transistors are modeled using predictive low-power 45 nm models [56]. Transmission-gate transistor widths were set to 540 nm for PMOS and 360 nm for NMOS. These larger widths were chosen such that the gate, when open, would have little impact on signal strength. A strong and weak inverter were used, with the strong inverter having widths of 270 nm for

$$\begin{bmatrix} 0 & 0 & 0 & 0 & 0 & 0 \end{bmatrix} \begin{bmatrix} 1 & 1 & 1 & 1 & 0 & 0 \end{bmatrix} used$$

$$\begin{bmatrix} 1 & 2 & 3 & 4 & 5 & 6 \\ 2 & 3 & 4 & 5 & 6 & 1 \\ 3 & 4 & 5 & 6 & 1 & 2 \\ 4 & 5 & 6 & 1 & 2 & 3 \\ 5 & 6 & 1 & 2 & 3 & 4 \\ 6 & 1 & 2 & 3 & 4 & 5 \end{bmatrix} \begin{bmatrix} 1 & 2 & 3 & 4 & 5 & 6 \\ - & - & - & 5 & 6 & 1 \\ - & - & 5 & 6 & 1 & 2 \\ - & 5 & 6 & - & 2 & 3 \\ 5 & 6 & - & - & 3 & 4 \\ 6 & - & - & - & 4 & 5 \end{bmatrix} setMatrix$$

$$\begin{bmatrix} 2 & 1 & 1 & 1 & 1 & 1 \end{bmatrix} \begin{bmatrix} 2 & 2 & 1 & 1 & 2 & 1 \end{bmatrix} used$$

$$\begin{bmatrix} - & 2 & 3 & 4 & 5 & 6 \\ - & - & - & 5 & - & - \\ - & - & 5 & 6 & - & 2 \\ - & 5 & 6 & - & 2 & 3 \\ - & 6 & - & - & 3 & 4 \\ - & - & - & - & 4 & - \end{bmatrix} \begin{bmatrix} - & - & 3 & 4 & - & 6 \\ - & - & - & - & - & - \\ - & - & - & 6 & - & - \\ - & - & 6 & - & - & 3 \\ - & - & - & - & - & 4 \\ - & - & - & - & - & - \end{bmatrix} setMatrix$$

$$\begin{bmatrix} 2 & 2 & 2 & 1 & 2 & 2 \end{bmatrix} \begin{bmatrix} 2 & 2 & 2 & 2 & 2 & 2 \end{bmatrix} used$$

$$\begin{bmatrix} - & - & - & 4 & - & - \\ - & - & - & - & - & - \\ - & - & - & - & - & - \\ - & - & - & - & - & - \\ - & - & - & - & - & - \\ - & - & - & - & - & - \end{bmatrix} \begin{bmatrix} - & - & - & - & - & - \\ - & - & - & - & - & - \\ - & - & - & - & - & - \\ - & - & - & - & - & - \\ - & - & - & - & - & - \\ - & - & - & - & - & - \end{bmatrix} setMatrix$$

Fig. 4.22 A step through of **createTestGroups** with $T = 6$, $m = 1$, and $r = 4$

PMOS and 180 nm for NMOS, and the weak inverter having 135 nm for PMOS and 90 nm for NMOS. These were chosen such that the majority of transistor W/L ratios were 2/1 for NMOS and 3/1 for PMOS. The charge-sharing capacitor C_{charge} was modeled at 10 pF, chosen to be an order of magnitude larger than the fault-free capacitance of the TSV network. This is sufficiently large to achieve good resolution in measurement without being so large that charge times are unreasonable or leakage becomes a significant issue. The power supply voltage V_{dd} for both the probe electronics and the circuit under test was set at 1.2 V.

All test time reductions shown in this section refer to the reduction in test time compared to the case of testing each TSV individually. This test time reduction considers only the time needed to charge C_{charge}, and not the time needed for control signals or the movement of the probe card. A reduction of 0 % means that the algorithm could not determine a solution that resulted in test times lower than the sequential testing baseline case. To simplify the presentation, the resolution r is given as the maximum number of TSVs that can be tested in parallel.

Figure 4.23 lists the test time reduction versus the number of faulty TSVs that must be pinpointed, for different resolution values in a 20-TSV network. It can be seen that, in general, increasing m results in less reduction in test time. This is

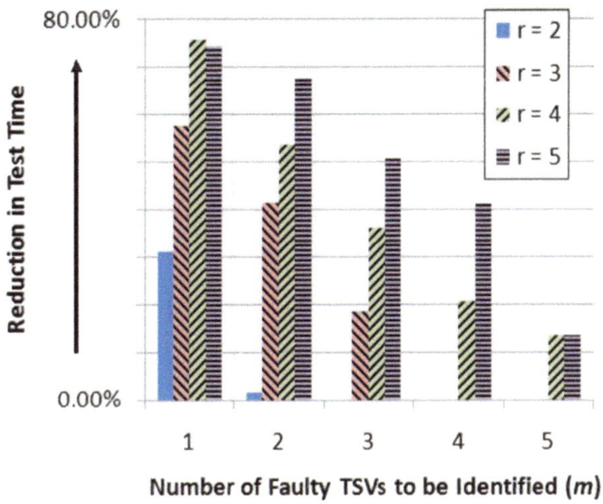

Fig. 4.23 Reduction in test time for a 20-TSV network

expected, as pinpointing more faulty TSVs requires more test groups. For the 20-TSV network, an increase in resolution tends to result in a decrease in test time. This is because there are enough TSVs in the network to capitalize on larger test groups. An exception occurs for a resolution of four TSVs with $m = 1$, which results in a larger reduction in test time than a resolution of five TSVs with $m = 1$. Although both of these optimizations produce ten test groups, a resolution of $r = 4$ creates ten test groups, each with four TSVs. For a resolution of $r = 5$, only six test groups contain five TSVs, with two test groups of three TSVs and two test groups of two TSVs. Overall, this results in higher test time. Our algorithm allows for the entire design space to be quickly traversed, allowing the pinpointing of optimal values of r, given a limit on the maximum allowable resolution, to minimize test time.

The above effect can be seen more clearly in networks of fewer TSVs. Figure 4.24 reproduces the data from Fig. 4.23 for an 8-TSV network. As can be seen, a resolution of $r = 4$ leads to significantly shorter test times when compared to $r = 5$ for $m = 1$. This is because for $r = 4$ the algorithm produces, on average, larger as well as fewer test groups than for $r = 5$. Due to adjustments that the optimization algorithm made to *curRes* on account of the small size of the TSV network, test groups for higher resolutions and larger values for m were the same. Compared to the data for a 20-TSV network, higher values of m were a larger portion of the TSVs in the network. Thus sequential testing was more effective for $m \geq 3$.

It is necessary to examine the effect of the number of TSVs in the network on test groups. Figure 4.25 shows the reduction in test time with respect to m at a fixed resolution of $r = 3$ and various values for T. For a given resolution, larger reductions in test time are achieved when the values of T and m are such that most test groups contain the maximum number of TSVs that can be tested in parallel.

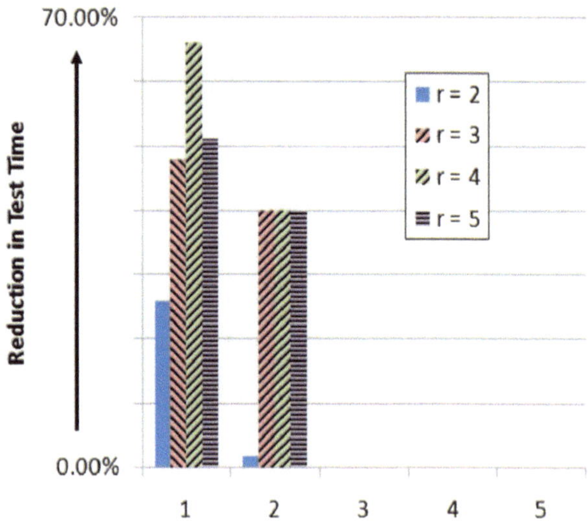

Fig. 4.24 Reduction in test time for a 8-TSV network

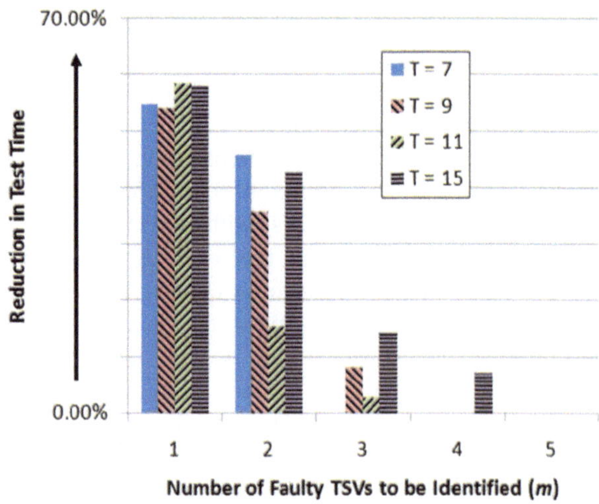

Fig. 4.25 Reduction in test time for resolution $r = 3$

For $m = 1$, this situation occurs for seven TSVs, i.e. $T = 7$. For $m = 2$ and $m = 3$, the greatest reduction in test time is obtained for 11 and 15 TSVs, respectively. These results further motivate the need for careful design and optimization of all parameters. Automated design tools can use the fast algorithm described in this section for this purpose. The generation of an array of data across values of T from 5 to 20, m from 1 to 5, and r from 2 to 5 took less than 3 s of CPU time.

Finally, the number of test groups produced during optimization is explored. Figure 4.26 shows the number of test groups produced with respect to m at a resolution of $r = 4$ and for various values of T. Data points are not shown for

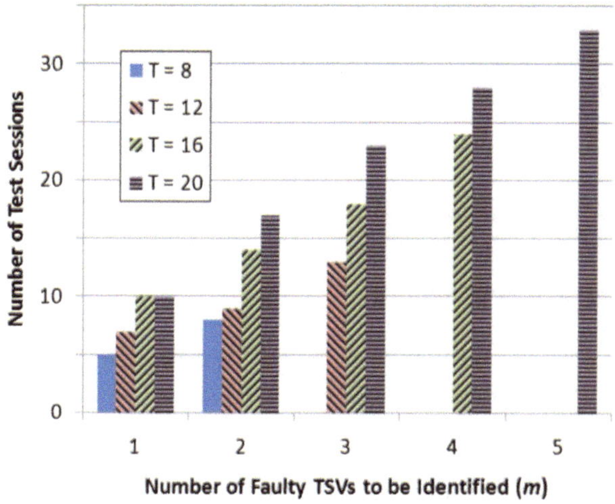

Fig. 4.26 Number of test groups produced for resolution $r = 4$

values of m and T for which the algorithm could not reduce test time relative to the baseline case of sequential TSV testing. For smaller values of m, the algorithm often produced fewer test groups when compared to the number of tests needed for each TSV individually. With larger TSV networks, it is possible to reduce test time while increasing the number of groups needed. This reduction in test time (but with more test groups) increases controller and routing complexity. It remains an open problem to determine the best trade-off by considering implementation cost for the test application scheme.

4.3.3 Limitations of the createTestGroups *Algorithm*

Though the **createTestGroups** can provide significant reductions in test time, it does not create optimal test groups in every design situation due largely to its greedy nature in assigning TSVs to test groups. Consider again the example of Fig. 4.22 where $T = 6$, $m = 1$, and $r = 4$. The algorithm begins by placing four TSVs in the test set $\{1, 2, 3, 4\}$. Unable to create any more 4-TSV test sets, it then produces the test set $\{1, 5, 6\}$. It once again is unable to create at test set at resolution $r = 3$, and so creates the test sets $\{2, 5\}$ and $\{3, 6\}$. Finally, it can only place the last TSV alone in the test set $\{4\}$. This does result in a considerable reduction in test time over serial TSV testing (44.40 %), but this is not an optimal test time.

Further reduction in test time could be achieved if the test sets $\{1, 2, 3\}$, $\{1, 4, 5\}$, $\{2, 4, 6\}$, and $\{3, 5, 6\}$ were used instead. These tests result in a significant 63.93 % reduction in test time over the serial test case, and a 35.11 % reduction over the test solution developed by the algorithm. It should be noted that the algorithm can produce the more optimal test set if its parameters are instead set to $T = 6$, $m = 1$, and $r = 3$, but this example does demonstrate the sub-optimality of the algorithm overall and how results can be improved if the algorithm is utilized as part of a larger optimization framework.

4.3.4 Summary

Section Summary

- Significant reductions in TSV test time can be achieved if multiple TSVs are tested simultaneously within the same TSV network.
- A limitation on the number of faulty TSVs within a TSV network that must be uniquely identified can be utilized to further reduce test time.
- An algorithmic approach to creating parallel test groups shows significant test time reductions over the case of serial TSV test within a TSV network, with reductions above 70 % in some examined cases.
- Test time reduction can be achieved in two ways—reduction in the number of tests required and/or reduction in the time needed to charge the charge-sharing capacitor for any given test.
- Due to the greedy nature of the algorithm, results are sub-optimal in some design cases.

4.4 Conclusions

This chapter has examined novel on-die DFT methods and measurement techniques combined with pin electronics that allow probing-based pre-bond testing of TSVs. It has been demonstrated how a probe card can be used together with the DFT architecture to measure resistance and capacitance, as well as to perform stuck-at and leakage tests. These parametric tests are applied to a network of TSVs, and HSPICE simulation results highlight the effectiveness of this approach. It is possible to test not only multiple TSV networks in parallel, but also several TSVs in each network in parallel if some loss of resolution in the measured data is acceptable. The test method yields reliable and accurate results even in the presence of process variations and multiple defective TSVs. The proposed method highlights the relevance of commercial testers and the role that tester companies can play in the maturation of 3D SIC test methods. It also demonstrates the need for cheap, effective low-force probe techniques to minimize touchdowns and damage to dies and TSVs as well as to keep the cost of probing reasonable when compared to BIST techniques.

Furthermore, the problem has been formulated of identifying faulty TSVs when using pre-bond probing to test TSVs within the same TSV network simultaneously. This problem can be described in terms of test time, resolution for fault detection, and the number of test groups required to localize a given number of defective TSVs. An efficient algorithm has been introduced for calculating test groups for parallel TSV testing within TSV networks. Results have been provided to highlight the significant reductions in test time achievable with parallel test. The test time reduction depends on the number of TSVs in a network, the number of faulty TSVs to detect, and the minimum resolution needed for measurements. The results highlight the need for a general multi-objective framework in which the proposed algorithm can be an important component.

Chapter 5
Pre-bond Scan Test Through TSV Probing

5.1 Introduction

Previous chapters have discussed the need for pre-bond KGD test to enable wafer matching, die binning, and other methods of ensuring stack yield. Chapters 3 and 4 presented methods through BIST and probing to enable pre-bond TSV test. While TSV test is important for KGD test, it covers only a small fraction of the tests that must be performed to achieve complete KGD test. In particular, the majority of die area is dedicated to logic and associated memory.

Chapter 2 examined several 3D memory architectures and a variety of BIST-based test methods, many of which can be utilized for pre-bond test. The majority of the remainder of die area after accounting for TSVs and memory is occupied by digital logic circuits that utilize the memory and TSVs. It is crucial that this logic is also tested during pre-bond KGD test.

Many test architectures are discussed in the literature [27, 35] to enable pre-bond test, including the die-level standard wrapper that is examined in Chap. 7. However, these architectures rely on the deposition of oversized probe pads on either or both of those TSV pillars or face-side TSV contacts that will be utilized for pre-bond logic test. These probe pads are sized to allow single contact of probe needles to a TSV. They require a significant amount of space and considerably limit TSV pitch and density. Therefore, only a limited number of probe pads are utilized for pre-bond test. This significantly limits the pre-bond test bandwidth available for logic test, increasing test time and cost.

To address the above challenges, this chapter explores a novel method for pre-bond testing of die logic through backside probing of thinned dies. It extends the test architecture discussed in Chap. 4. While the probing technique of Chap. 4 focused only on TSV test, this chapter focuses on scan-test of die logic utilizing scan chains that can be reconfigured for pre-bond test to allow scan-in and scan-out through TSVs. This method does not require many oversized probe pads, save for a few critical signals such as power, ground, and test/functional clocks. A significant benefit of the method outlined in this chapter is that, coupled with the architecture

B. Noia and K. Chakrabarty, *Design-for-Test and Test Optimization Techniques for TSV-based 3D Stacked ICs*, DOI 10.1007/978-3-319-02378-6_5,

described in Chap. 4, it enables both pre-bond TSV/microbump test as well as pre-bond structural test under a single test paradigm. Furthermore, probe pads do not limit the test bandwidth of pre-bond test, so a high-bandwidth pre-bond test can be performed quickly through the methods outlined in this chapter. Several different scan configurations are examined in this chapter, each providing varying degrees of test parallelism depending on design constraints. A variety of simulations will be discussed that demonstrate the feasibility of scan test through TSV probing, including area overhead, power/current delivery needs, current density in TSVs, and scan clock frequencies.

The rest of this chapter is organized as follows. Section 5.2 examines the architecture of Chap. 4 in context of pre-bond scan test, and introduces two new kinds of GSFs depending on the type of TSV they are driving. Section 5.2.1 introduces the proposed scan architecture and test method for performing pre-bond scan test. Section 5.2.2 presents HSPICE simulation results for a number of dies with TSVs across two logic-on-logic 3D benchmarks, highlighting the feasibility of the method presented discussed in this chapter. Finally, Sect. 5.3 concludes the chapter.

5.2 Pre-bond Scan Test Through TSV Probing

In Chap. 4, a measurement and DFT technique was introduced to enable the pre-bond test of TSVs through probing. This method utilized a die wrapper similar to that discussed in Chap. 7 but replaced the boundary scan flops with gated scan flops (GSFs). For pre-bond scan test, the directionality of TSVs in a TSV network is important, so it is necessary to distinguish between GSFs on *sending* and *receiving* TSVs. A sending TSV is a TSV that is driven by logic on its own die during functional operation and sends a signal to another die. A receiving TSV is a TSV that is driven by logic on another die during functional operation and receives a signal.

Figure 5.1 shows the gate-level design of a bidirectional GSF. The receiving path of the GSF is highlighted with an arrow. As discussed in Chap. 4, a GSF multiplexes between a test input and a functional input and can be connected to other GSFs to form a scan chain. The difference is that the GSFs include a buffer of two inverters and a transmission gate at the output of the flop, which accepts an 'open' signal to switch between a low- and a high-impedance output. This design effectively allows the TSV to be driven by the GSF or to be left floating. GSFs on receiving TSVs must be bidirectional GSFs because the GSF must be able to drive the TSV during pre-bond TSV test.

In Chap. 4, the GSFs were included before each TSV to enable pre-bond probing of TSVs. It was shown that by using probe needles larger than an individual TSV, groups of TSVs can be deliberately shorted together to form a single circuit called a TSV network. Using the GSFs, the resistance of each TSV can be accurately

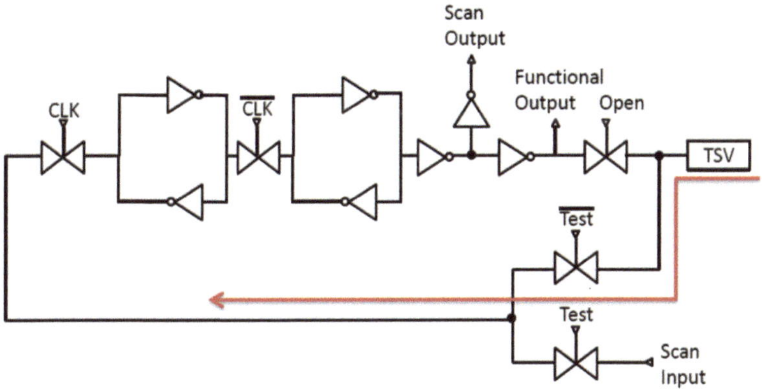

Fig. 5.1 An example gate-level design for a bidirectional gated scan flop with receiving path highlighted

determined, along with the average capacitance of each TSV. Contact force and variations in contact quality between TSVs were shown to have little effect on the ability to accurately characterize TSVs.

To enable pre-bond scan test using the same architecture, scan chains are reconfigured into a pre-bond test mode in which scan inputs and scan outputs are connected to TSV networks. This allows the probe station to apply test patterns to the die and to read test responses through the scan chains and pre-bond TSV scan I/O. A key advantage of using TSVs for pre-bond scan test is that not all TSVs need to be contacted for die logic test. It is necessary to contact only those TSVs that are required for pre-bond scan. Results for a 3D benchmark that will be discussed further in Sect. 5.2.2 show that for 100 scan chains for pre-bond test, as few as 10.7% of the TSVs need to be contacted. Therefore, only one touchdown is likely needed for pre-bond scan test, and this can be the second touchdown required for pre-bond TSV test to allow for scan test after all TSVs have been tested to be fault free.

5.2.1 *Performing Pre-bond Scan Test Through TSV Probing*

This section describes the test architecture and methods required to perform pre-bond scan test. A post-bond scan architecture is assumed that is compatible with the die wrappers discussed in Chap. 7, as shown in Fig. 5.2. Figure 5.2a shows a single scan chain and a number of boundary scan flops. The scan chain consists of typical scan flops (SFs), while boundary scan registers at the TSV interface are GSFs. As with die wrappers, some landing pads must be supplied for providing essential signals to the die, such as power, ground, and clocks. The post-bond scan input and scan output for a scan chain enter the die through the boundary register. In the bottom die in a stack, this interface is through external test pins or a JTAG

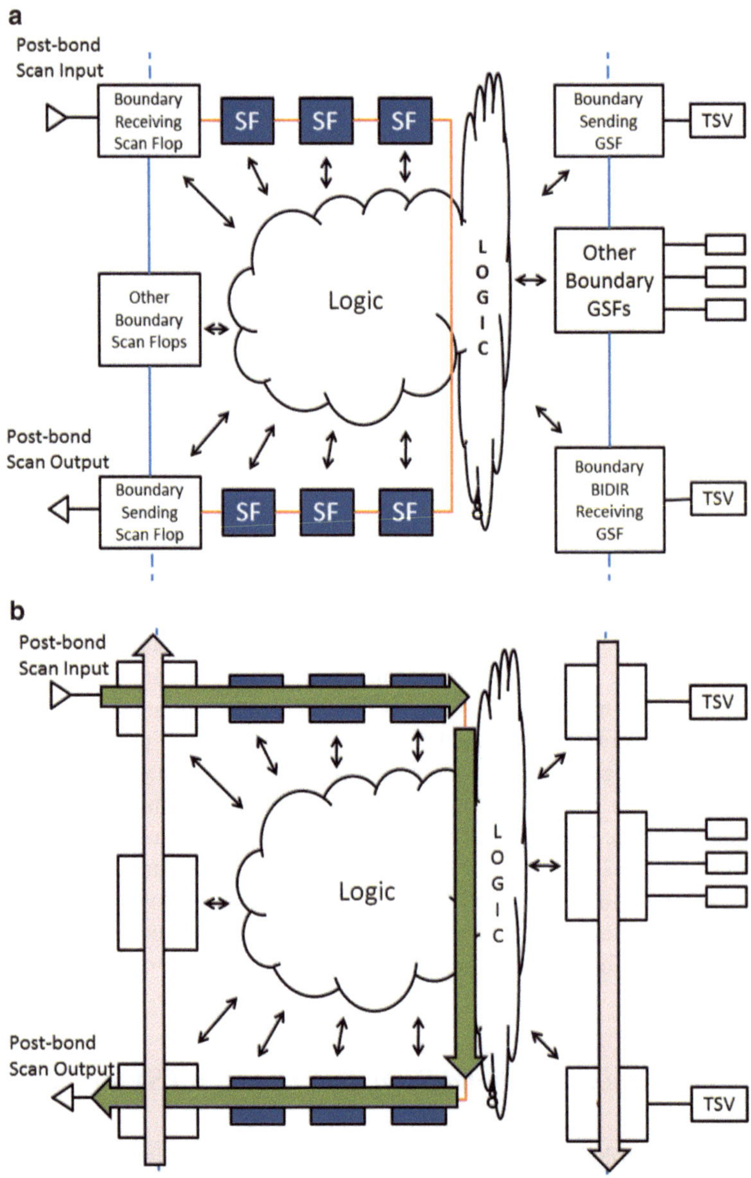

Fig. 5.2 The assumed post-bond scan architecture: (**a**) scan chains and logic; (**b**) movement of test data

test-access port. For other dies in the stack, scan I/Os are connected to the dies below them in the stack. Parallel loading of the boundary registers decreases test time, but serial scan test is also available by shifting through the boundary scan chain. This is illustrated in Fig. 5.2b, which shows the post-bond movement of test data. Test data can be shifted not only through the internal scan chain, but also around the boundary registers. All scan flops interact with die logic.

Multiplexers are added to the scan path to allow scan chains to be reconfigured to a pre-bond mode in which their scan-in and scan-out connections are through TSVs, as shown in Fig. 5.3a. A receiving GSF is chosen for the reconfigured scan-in and a sending GSF is chosen for the scan-out. Because many boundary scan registers are logically separated from internal scan chains in the post-bond mode, they need to be stitched to the scan path in pre-bond mode to enable testing. Multiplexers are added in as few places as possible to achieve access to all internal and boundary scan flops in order to minimize hardware overhead.

Consider the multiplexers added to a single scan chain in Fig. 5.3a. The receiving GSF, which now acts as the pre-bond scan input, is enabled to accept its functional input driven through the TSV. Its scan output is then multiplexed into the boundary scan chain. This is done such that the sending GSF used as a pre-bond scan output and the receiving GSF used as a pre-bond scan input will interface with scan flops that are adjacent to one another in the post-bond scan chain. The output of the boundary scan flop that is used to feed the pre-bond scan input is then multiplexed into the scan chain. The post-bond scan output, post-bond scan input, and other boundary registers are stitched into the scan chain. Finally, the sending GSF used as a pre-bond scan output is multiplexed to the end of the scan chain. The pre-bond movement of test data is shown in Fig. 5.3b. The combinational logic is not shown so as to retain clarity; it is the same as in Fig. 5.3a. Arrow color changes in the figure so as not to confuse the overlapping arrows.

The reconfigured pre-bond scan chain in Fig. 5.3 demonstrates one of several possible pre-bond scan configurations (Configuration A). In this example, the pre-bond scan chain's scan-in and scan-out terminals are part of the same TSV network. Under these conditions, the scanning in of test data and the scanning out of test responses must be done separately. This is because, in order to scan in test data, the transmission gate on the receiving GSF must be set to its low-impedance state while all other gates must be set to their high-impedance states. Likewise, while scanning out, the sending GSF's gate must be set to low impedance while all others are set to high impedance. Because scan-in and scan-out occur on the same network, the maximum number of scan chains that can be tested in a single touchdown is equal to the number of TSV networks formed. In other words, the number of scan chains can at most be equal to the number of probe needles. Furthermore, if current or power limits cause the maximum scan clock frequency to be different for scan input and scan output, then the appropriate frequency must be used for the corresponding operation.

A second possible pre-bond scan configuration (Configuration B) involves the scan input and scan output on separate TSV networks, an example of which is shown in Fig. 5.4. In this case, test responses can be scanned out while test patterns are scanned in. The maximum number of scan chains that can be tested per touchdown is reduced to half of the number of probe needles (or half of the number of TSV networks). Both scan input and scan output operations must occur at the lower of the possible scan frequencies, because both operations occur simultaneously.

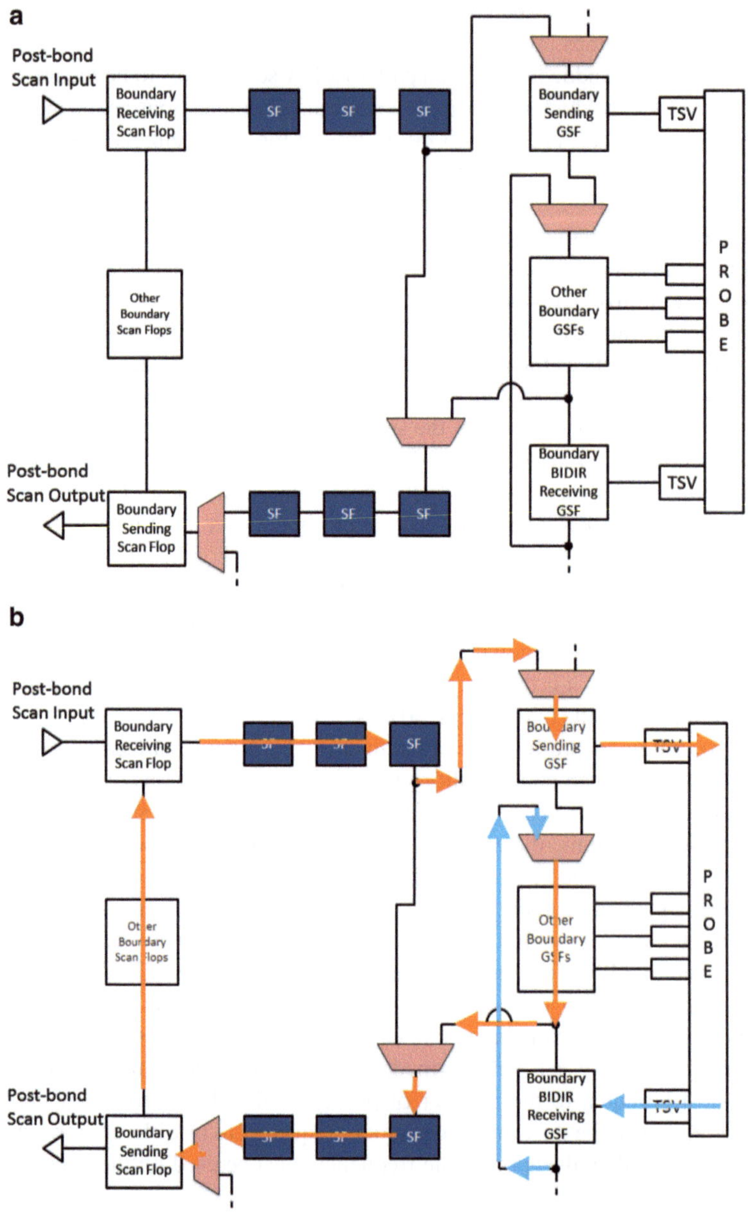

Fig. 5.3 Reconfigurable scan chains for pre-bond test: (**a**) added multiplexers; (**b**) movement of test data

It should be noted that pre-bond functional test cannot be conducted while using TSV networks, because it is not possible to supply individual inputs to TSVs within a network at the same time.

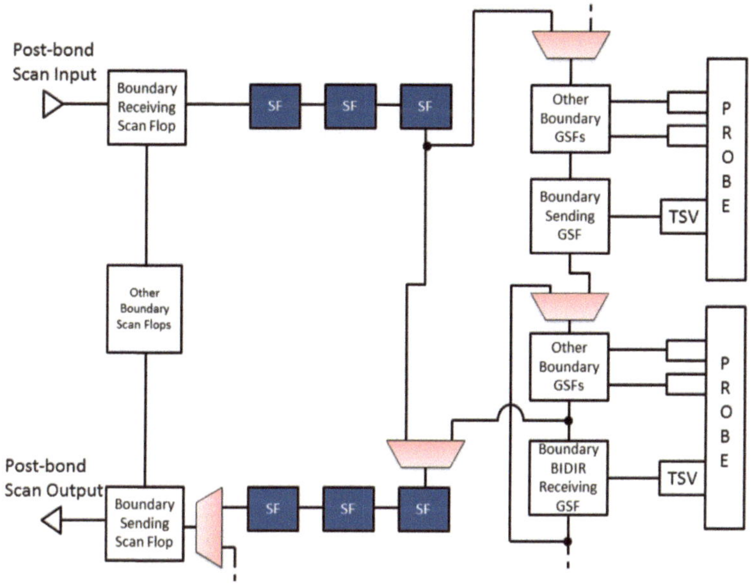

Fig. 5.4 A reconfigurable scan chain with pre-bond scan input and scan output on different TSV networks

Pre-bond scan configurations can also be designed such that two or more scan inputs and/or scan outputs belong to the same TSV network. Such a configuration is desirable in a number of scenarios. Design constraints such as routing complexity or layout difficulty may prevent the routing of a scan chain's pre-bond I/O to an independent TSV network. In such a case, the scan chain may be required to share a pre-bond scan input, output, or both, with TSV networks that already have pre-bond scan I/O routed to them. In another scenario, there may exist more post-bond scan chains than there are pre-bond TSV networks in a single touchdown. Because realigning the probe card and performing a second touchdown significantly increases test time, it is preferable to test all scan chains in a single touchdown. In this case, sharing TSV networks between pre-bond scan I/O can result in test times shorter than if two scan chains are stitched together to form a single, longer scan chain.

Figure 5.5 shows a pair of examples where two separate scan chains share TSV networks. In Fig. 5.5a, the pre-bond scan inputs and outputs of both scan chains are routed to the same TSV network (Configuration C). In Fig. 5.5b, reconfigurable scan chains 1 and 2 share a TSV network for their pre-bond scan inputs, but have independent TSV networks for their scan outputs (Configuration D). When scan chains share a TSV network across their pre-bond scan inputs, patterns can be applied using a broadcast method to reduce test time. During the broadcast of test patterns, the scan chains must receive unique shift signals such that one or both can shift in bits depending on which bit is applied to the TSV network. Test patterns

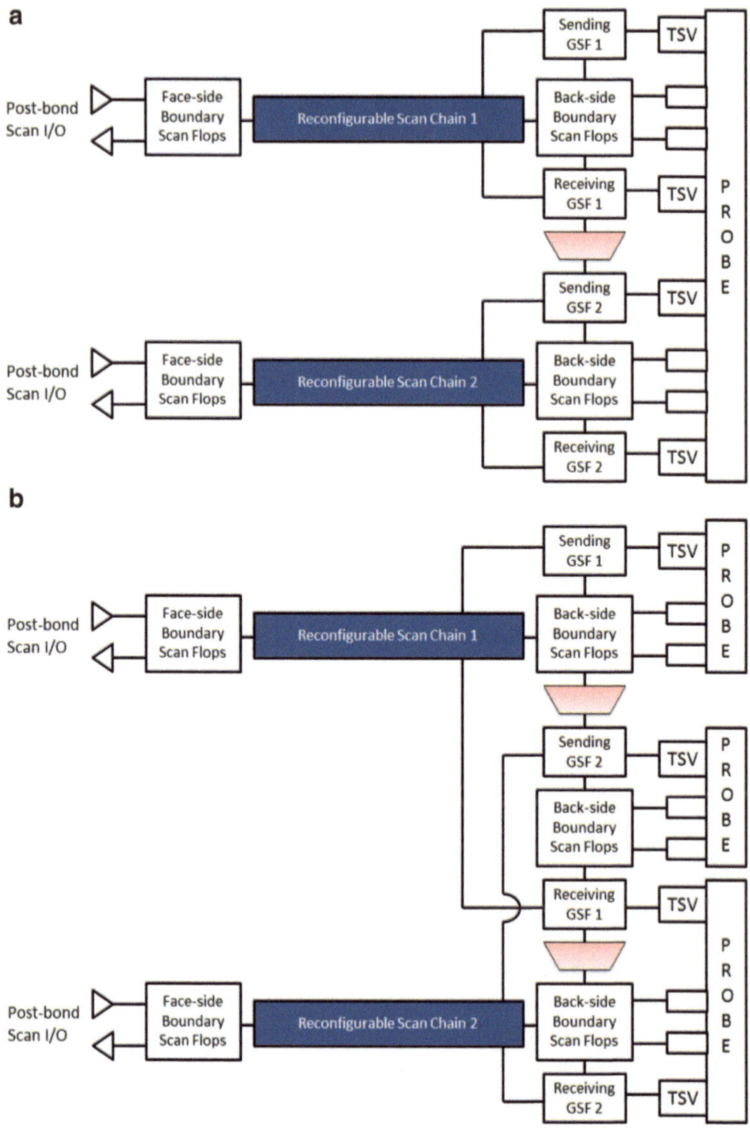

Fig. 5.5 Reconfigurable scan chains with two pre-bond scan inputs on the same TSV network and (**a**) scan outputs on the same TSV network or (**b**) scan outputs on separate TSV networks

for both scan chains can then be combined into single patterns that require fewer test clock cycles to scan in than scanning in the patterns serially. When scan outputs share a TSV network, test responses must be scanned out serially. Therefore, the configuration of Fig. 5.5a must utilize either a serial or broadcast scan-in and a serial scan-out, and scan-in and scan-out operations cannot occur simultaneously.

For the configuration of Fig. 5.5b, scan-in must occur serially or through broadcast, but scan-out from the two scan chains can occur in parallel. Scan-in and scan-out operations can occur simultaneously.

From a test time perspective, it can be determined which configuration is best to use for a given design. The following design constraints are utilized:

- s - The number of pre-bond scan chains created during reconfiguration.
- p - The number of patterns to be applied during pre-bond scan test.
- m - The number of scan cells in the longest pre-bond scan chain. This value is assumed to be constant across touchdowns for determining test times in this chapter, though it need not be.
- l_i - The number of bits in the length of the ith pattern, where $i = 0$ is the first pattern of a pattern set. This variable is only required for configurations 3 and 4, which utilize broadcast patterns. Thus, each pattern can be of a varying length and will generally be larger than m.
- n - The number of probe needles available for TSV networks on the probe card.
- t - The time required for the alignment and touchdown of the probe card.
- f_{in} - The maximum scan-in clock frequency that can be used.
- f_{out} - The maximum scan-out clock frequency that can be used.

The objective is to determine an equation for the test time T required for each configuration given the constraints above and choose the configuration that results in the shortest test time. Because the time required for alignment and touchdown is generally much longer than the time required to perform scan tests, it is often better to use configurations that require only a single touchdown.

For configuration A, scan-in and scan-out operations occur sequentially because scan I/O utilize a shared TSV network and probe needle. To speed this process, scan-in operations can use the maximum scan-in frequency f_{in} and scan-out operations can use the maximum scan-out frequency f_{out}. The equation for the test time for Configuration A is:

$$T_A = \left\lceil \frac{s}{n} \right\rceil \cdot \left(\frac{m \cdot p}{f_{in}} + \frac{m \cdot p}{f_{out}} \right) + \left\lceil \frac{s}{n} \right\rceil \cdot t \tag{5.1}$$

The number of touchdowns required to perform pre-bond scan test is given by $\lceil \frac{s}{n} \rceil$. This is then multiplied by the time required to apply all the test patterns and receive all test responses for each touchdown and added to the time required to perform all alignment and touchdown operations ($\lceil \frac{s}{n} \rceil \cdot t$).

For Configuration B, scan-in and scan-out operations can occur in parallel, which reduces the time required to apply patterns and receive test responses when compared to the time required by Configuration A. However, Configuration B can interface with half-as-many scan chains per touchdown as Configuration A. The test time for Configuration B is written as:

$$T_B = \left\lceil \frac{2s}{n} \right\rceil \cdot \left(\frac{m \cdot (p+1)}{\min\{f_{in}, f_{out}\}} \right) + \left\lceil \frac{s}{2n} \right\rceil \cdot t \tag{5.2}$$

Configuration C allows for significant consolidation of scan chains across TSV networks, allowing for the test of twice as many scan chains per touchdown as Configuration A and four times as many as Configuration B. Scan-in and scan-out operations are performed sequentially, requiring two scan-out cycles for each scan-in cycle due to the need to scan out two scan chains worth of responses for each TSV network. Furthermore, patterns are of variable length due to the compression required to generate broadcast patterns. The test time for Configuration C is thus calculated as:

$$T_C = \left\lceil \frac{s}{2n} \right\rceil \cdot \left(\frac{\sum_{i=0}^{p} l_i \cdot p}{f_{in}} + 2 \cdot \frac{\sum_{i=0}^{p} l_i \cdot p}{f_{out}} \right) + \left\lceil \frac{s}{2n} \right\rceil \cdot t \tag{5.3}$$

Lastly, Configuration D allows for parallel scan-in and scan-out operations while allowing for the test of more scan chains per touchdown than Configuration B. It utilizes the broadcast pattern set, and its test time is found to be:

$$T_D = \left\lceil \frac{\frac{3}{2}s}{n} \right\rceil \cdot \left(\frac{\sum_{i=0}^{p} l_i \cdot (p+1)}{\min\{f_{in}, f_{out}\}} \right) + \left\lceil \frac{\frac{3}{2}s}{n} \right\rceil \cdot t \tag{5.4}$$

Though these equations can act as a guide in determining which configuration to use for a design, they only encompass test time considerations for creating the reconfiguration architecture. In reality, design and technology constraints, such as routing complexity, area overhead, and so forth, will also influence which configurations are feasible for a given design.

Figure 5.6 shows the test times of Configurations A, B, C, and D while varying the number of probe needles available for TSV network creation from 10 to 200. Parameter values were chosen to show the difference between the configurations, with s of 50, p of 1,000, m of 300, each l_i value set to 400, f_{in} of 150 MHz, and f_{out} of 100 MHz. The alignment and touchdown time t for Fig. 5.6a is 1.5 ms, which is relatively fast but is used to ensure that the test times of the die are not eclipsed by t to provide a complete picture of the various configurations. Figure 5.6b is produced when $t = 100$ ms, which is significantly longer than the time required to perform structural test using the given parameters, and provides a realistic look at the differences between the configurations in practice.

As Fig. 5.6a demonstrates, there is pareto-optimality among the configurations with regard to n. For low values of n, Configurations C and A tend to result in lower test times as they provide a higher compression ratio for scan chains among TSV networks. At higher values of n, Configurations B and D result in lower test times as they provide higher test parallelism with regard to pattern application. Figure 5.6b shows these effects in a more realistic environment, where utilizing the configuration that best matches the pre-bond scan test bandwidth and utilizes only a single touchdown has the greatest impact on test time. Which configuration will provide the lowest test times depends on both the design parameters and the probe card constraints. Section 5.2.2 will explore this issue further with results from benchmark circuits.

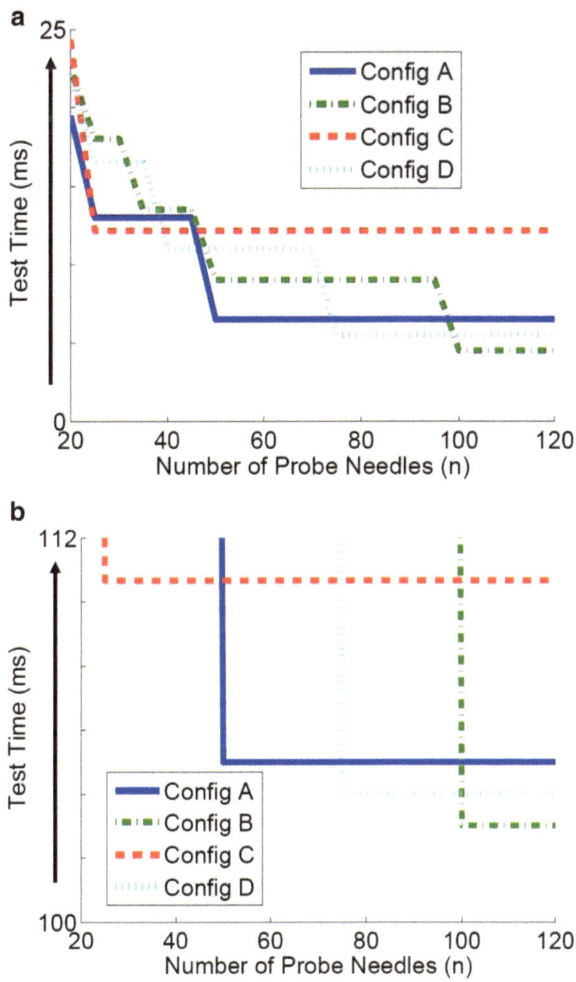

Fig. 5.6 The test times of Configurations A, B, C, and D with varying numbers of probe needles and alignment and touchdown time of (**a**) $t = 1.5$ ms or (**b**) $t = 100$ ms

5.2.2 Feasibility and Results for Pre-bond Scan Test

This section addresses a number of key criteria needed to demonstrate the feasibility of the proposed method:

- The current needed to be delivered to the device under test during pre-bond scan test must fall within the current-carrying capacities of TSVs and probe needles.
- The speed at which highly capacitive TSV networks are charged and discharged must be reasonable such that the pre-bond scan test time is low.
- The area overhead of the proposed method must be small.
- That boundary scan registers are necessary to achieve high coverage in pre-bond scan test.

Simulation results are presented demonstrating the feasibility of the methods presented in this chapter. Simulations were conducted in HSPICE on two 3D logic-on-logic benchmarks. The resistance and capacitance used for each TSV of 5 μm diameter were 1 Ω and 20 fF, respectively [26, 53]. Transistors were modeled using a predictive low-power 45 nm model [56] except where otherwise noted. Transmission-gate transistor widths were set to 540 nm for PMOS and 360 nm for NMOS. These larger widths were chosen such that the gate, when open, would have little impact on signal strength. For each GSF, a strong and weak inverter were used, with the strong inverter having widths of 270 nm for PMOS and 180 nm for NMOS, and the weak inverter having 135 nm for PMOS and 90 nm for NMOS. These were chosen such that the majority of transistor W/L ratios were 2/1 for NMOS and 3/1 for PMOS. The power supply voltage for both the probe and the circuit was taken to be 1.2 V.

5.2.2.1 3D IC Benchmarks

Because 3D IC benchmarks are not available in the public domain, two benchmarks were created from cores available through the OpenCores set of benchmarks [85]. A Fast Fourier Transform (FFT) circuit and a Reconfigurable Computing Array (RCA) circuit were utilized. Both are synthesized using the Nangate open cell library [101] at the 45 nm technology node [56]. The total gate count after synthesis is 299,273 with 19,962 flip-flops for the FFT circuit, and 136,144 gates with 20,480 flip-flops for the RCA circuit. Both designs were partitioned into 4 dies, with the gate counts in each die of the FFT stack being 78,752, 71,250, 78,367, and 70,904, respectively. For the RCA stack, gate counts for each die were 35,500, 34,982, 32,822, and 32,840, respectively. The logic gates in each die are placed using Cadence Encounter, and TSVs are inserted in a regular fashion, using a minimum spanning tree approach [86]. Back-to-face bonding is assumed, which means that TSVs are present only in the first three dies. The TSV counts for each die in the FFT stack are 936, 463, and 701, respectively, and for the RCA stack are 678, 382, and 394, respectively. The TSV diameters are 5 μm. The circuits were routed such that each TSV has a small microbump sized at 7 μm, and the total TSV cell size including the keep out zone is 8.4 μm, which corresponds to six standard cell rows. Each die is then routed separately in Cadence Encounter. The bottom die of the FFT 4-die layout is shown in Fig. 5.7, with TSVs in white and standard cells in green.

Boundary scan cells were added at the TSV interface. The need for inserting boundary registers at the TSV interface can be motivated by examining die 0 of the 4-die FFT benchmark. Without boundary scan registers, the pre-bond stuck-at fault coverage is only 44.76 %. With boundary registers added, the coverage increases to 99.97 % for stuck-at test patterns and 97.65 % for transition test patterns. This is a significant increase, especially considering that the die only contains 936 TSVs, and an industry design may contain tens of thousands of TSVs.

The area overhead of the boundary scan GSFs and scan chain reconfiguration circuits is shown for the FFT dies with TSVs in Table 5.1 and for the RCA dies with

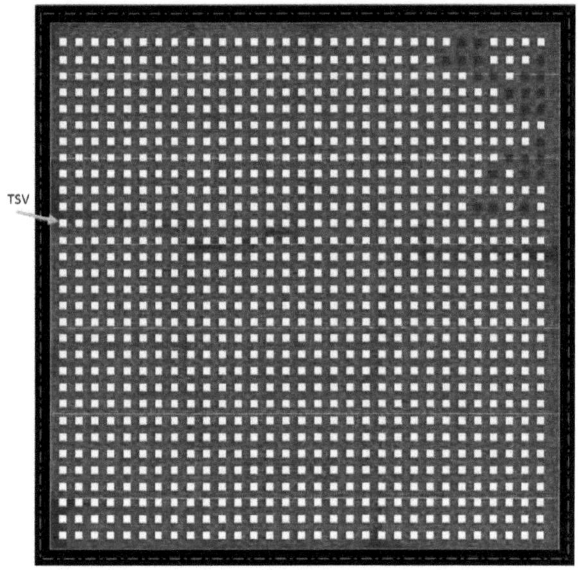

Fig. 5.7 Layout of die 0 of the 4-die FFT benchmark, with standard cells in *green* and TSVs in *white*

Table 5.1 A comparison of the worst-case results of three dies with TSVs in the FFT 3D stack

Test parameter (FFT stack)	Die 0	Die 1	Die 2
Peak current	1 mA	1 mA	1.1 mA
Average current (stuck-at)	300 μA	294 μA	327 μA
Average shift current (transition)	387 μA	300 μA	335 μA
Average capture current (transition)	432 μA	341 μA	383 μA
Area overhead	2.2%	1.0%	1.2%

TSVs in Table 5.2. These results show area overheads between 1.0 % and 2.9 % of the total number of gates. Generally, the area overhead was higher for the RCA benchmark because the benchmark contains significantly fewer gates per die than the FFT benchmark, while at the same time containing nearly as many flops and without a significant reduction in TSVs. This means that many boundary scan cells need to be added to the RCA dies, and there is a similar number of scan chains that need reconfiguration circuitry between the two benchmarks.

5.2.2.2 Simulation Results

The feasibility of performing scan test through probe needles is first examined in terms of sourcing and sinking currents. To determine an upper limit on the current drawn, scan chains were inserted into the benchmark. In order to manage the

Table 5.2 A comparison of the worst-case results of three dies with TSVs in the RCA 3D stack

Test parameter (RCA stack)	Die 0	Die 1	Die 2
Peak current	0.8 mA	0.8 mA	0.8 mA
Average current (stuck-at)	279 μA	288 μA	242 μA
Average shift current (transition)	287 μA	321 μA	261 μA
Average capture current (transition)	327 μA	331 μA	300 μA
Area overhead	2.9 %	1.7 %	1.9 %

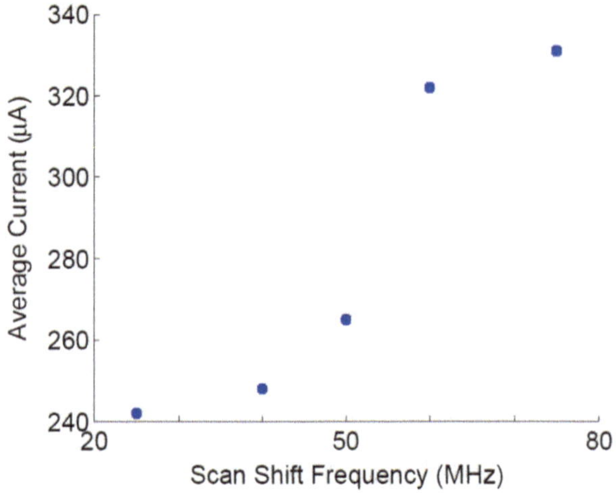

Fig. 5.8 Average current drawn at 25, 40, 50, 60, and 75 MHz scan shift frequency for Die 0 of the FFT benchmark

complexity of circuit-level HSPICE simulation, scan chains were limited to a length of 8 (six internal scan cells and two boundary scan cells for pre-bond scan I/O per chain) for each die in each benchmark. Stuck-at and transition test patterns for this design were generated using a commercial ATPG tool and ordered based on toggle activity. Test generation yielded the toggle activity for each test pattern. For each die, two scan chains and associated logic were extracted for simulation in HSPICE based on toggle activity for the highest activity pattern and for an average activity pattern in the generated pattern set. By associated logic, it is meant that fan-in and fan-out gates for the scan cells of that scan chain were simulated in HSPICE up to primary I/O or other flip-flops. For the pattern with highest peak toggle activity, the scan chain and associated logic that yielded the largest number of transitions for that pattern were simulated. For the average pattern, a scan chain and associated logic based on the average number of toggling bits was simulated.

Figure 5.8 shows the current drawn for shifting in the highest-power stuck-at pattern for the worst-case scan chain and shifting out test responses at 25, 40, 50, 60, and 75 MHz shift frequency. The figure gives data for Die 0 of the FFT circuit.

Table 5.3 A comparison of the average-case results of three dies with TSVs in the FFT 3D stack

Test parameter (FFT stack)	Die 0	Die 1	Die 2
Peak current	1 mA	1 mA	1 mA
Average current (stuck-at)	289 μA	274 μA	291 μA
Average shift current (transition)	370 μA	281 μA	296 μA
Average capture current (transition)	412 μA	305 μA	344 μA

Table 5.4 A comparison of the average-case results of three dies with TSVs in the RCA 3D stack

Test parameter (RCA stack)	Die 0	Die 1	Die 2
Peak current	0.8 mA	0.7 mA	0.7 mA
Average current (stuck-at)	270 μA	270 μA	241 μA
Average shift current (transition)	277 μA	291 μA	246 μA
Average capture current (transition)	298 μA	317 μA	261 μA

At 50 MHz, current drawn averaged at around 300 μA and, at all frequencies, peaked at almost 1 mA for about a tenth of a nanosecond. For a high toggle-activity transition fault pattern using launch-off-shift and a 1 GHz functional clock for the same die, an average current of 432 μA is drawn during capture and peak current similar to that of stuck-at patterns.

Table 5.1 shows the peak current and average currents drawn for the worse-case stuck-at and transition pattern for the three dies with TSVs in the FFT benchmark. For transition test, an average current draw is shown both for the shift and capture cycles, with all scan-in shift cycles averaged together. Only a single peak current is shown because the results were nearly identical for stuck-at and transition patterns. Table 5.2 shows the same results for the first three dies of the RCA benchmark. The simulations were performed with a scan-shift frequency of 50 MHz and a functional clock of 1 GHz. Because neither the driver strength nor the TSV network size were changed for these simulations, maximum scan-in and scan-out frequencies were equal for the dies. Tables 5.3 and 5.4 show the same results for the average scan-chain and test pattern.

As is evident from the tables, the highest worst-case average current drawn for the stuck-at pattern was 327 μA, as experienced by Die 2 of the FFT benchmark. For the transition pattern, the highest worst-case current was 432 μA in Die 0 of the FFT benchmark during capture These worst-case currents were significantly less for the RCA benchmark, reaching as high as 288 μA for the stuck-at pattern and 331 μA during capture for the transition pattern. The average currents for the average scan chains and patterns were lower, as expected, though there is little change in the peak current draw.

Table 5.5 Low-power pattern generation results for Die 0 of the 4-die RCA benchmark

Activity target (RCA stack) (%)	Pattern inflation (%)	Coverage loss (%)	Peak current (mA)
90	1.4	0	0.75
80	3.9	0	0.71
70	8.1	0	0.66
60	16.3	3.5	0.58

It has been reported in the literature that a TSV can handle a current density higher than 70,000 A/cm^2 [87]. Published work on TSV reliability screening indicates that a sustained current density of 15,000 A/cm^2 is possible through a TSV without damage [88]. To sustain a peak current of 1 mA through a single 5 μm TSV in the pre-bond test method would require the TSV to be capable of handling a current density of 5,093 A/cm^2. To handle a 300 μA average current, a TSV must be capable of sustaining a current density of 1,528 A/cm^2. Both these numbers are well below the maximum allowable current density.

In addition to the current density limits of the TSVs, it is important to consider the amount of current that the probe needles can deliver. It has been shown in the literature that a 3 mil (76.2 μm) cantilever probe tip is capable of supplying 3 A of current for a short pulse time (less than 10 ms) [59, 89]. In the worst case, assuming that all scan chains and logic in the FFT benchmark draw the peak current at once, the probe tip would have to supply 3 A of current for less than 0.1 ns. This falls within the probe current-supply specification. If current supply from the probe is an issue, a variety of well-known methods can reduce peak and average test power on die during test, including partitioning the circuit into separate test modules, clock gating, and low-power patterns [90–92].

Table 5.5 shows the results of low-power pattern generation for Die 0 of the 4-die RCA benchmark in order to reduce test power. Column one shows the target peak toggle activity as a percentage of the unconstrained worst-case pattern toggle activity. Column two and column three provide the increase in pattern count as a percentage of the unconstrained pattern count and reduction in coverage as a percentage of the unconstrained coverage, respectively. Column 4 gives the peak current draw for the worst-case pattern using the same scan chain from the simulations from Table 5.4. As can be seen from the table, peak toggle activity can be significantly reduced (to roughly 70 % of the unconstrained toggle activity, resulting in a 0.66 mA peak current) without a loss of coverage and with at worst 8.1 % additional patterns. A reduction to 60 % results in some coverage loss (3.5 %), but can reduce peak current draw to 0.58 mA. These results demonstrate how low-power pattern generation can be used to reduce test power if it exceeds TSV or probe constraints.

Figures 5.9 and 5.10 show the average stuck-at current for Die 0 of the 4-die FFT benchmark with regard to changing TSV resistance and capacitance, respectively. Increases in TSV resistance resulted in an almost negligible increase in current draw, with a high TSV resistance of 5 Ω resulting in only a 0.13 % increase over the baseline current draw of 300 μA. As seen in Fig. 5.10, an increase in TSV

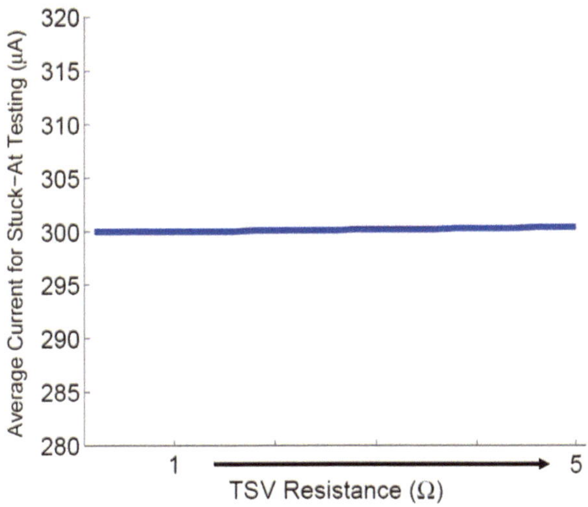

Fig. 5.9 Change in average stuck-at current versus TSV resistance

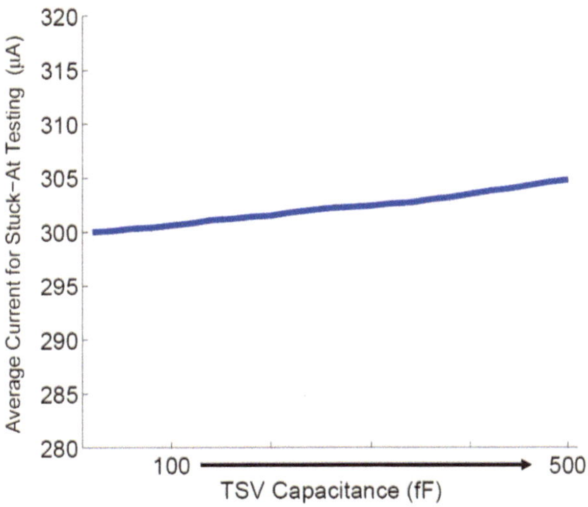

Fig. 5.10 Change in average stuck-at current versus TSV capacitance

capacitance had a slightly greater, though still minor, effect on stuck-at current draw, with a 500 fF TSV capacitance resulting in a 1.6 % increase over the baseline current draw. These results indicate that power consumption during test is dominated by die logic and not the TSVs. The current draw is greater for capacitance increases than resistance increases because the capacitance of all TSVs in a TSV network impacts test, due to an increase in net capacitance, whereas only the resistance of TSVs in use in the TSV network impacts test.

The feasibility of the proposed method from a test-time perspective can also be discussed. The frequency at which scan-in and scan-out can take place depends on a number of factors. Scan-in speed depends on the strength of the probe-needle driver, while scan-out speed depends on the strength of the TSV driver in the

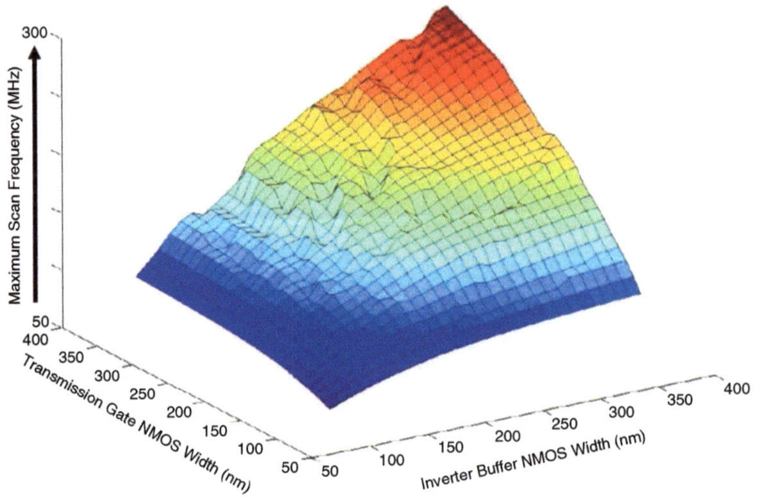

Fig. 5.11 Maximum scan-out frequency in a 11 TSV network with varying driver and transmission gate width for a 45 nm technology

sending GSF used as a scan output, which is the twin-inverter buffer shown in Fig. 5.1, and the width of the GSF transmission gate. Both the GSF driver and the probe driver must be able to charge and discharge the TSV network capacitance quickly enough to meet the setup and hold times of the scan flops given the test clock frequency. Therefore, the number and capacitance of TSVs in a network also influence maximum scan clock frequency.

Scan frequency simulations were performed on Die 0 of the FFT benchmark assuming a probe card with 100 probe needles [76]. The design contains 936 TSVs and it is assumed that TSV networks are roughly balanced, so a worse-case network of 11 TSVs was simulated. This results in a network capacitance of 220 fF. Simulations were performed using both the 45 nm low-power technology model and a 32 nm low-power technology model. It is assumed that drivers in the probe needle can be significantly stronger than drivers on the die itself, so simulations were performed only for the scan-out frequency as this would be the limiting factor to test time. The widths of the inverter driver and the transmission gate were varied and a maximum scan frequency was calculated by measuring the amount of time required for a rising or falling signal to charge or discharge 75 % or 25 % of Vdd, respectively. The results of these simulations are shown in Fig. 5.11 for the 45 nm technology node and in Fig. 5.12 for the 32 nm technology node.

As can be seen from Figs. 5.11 and 5.12, maximum scan-out frequency depends strongly on both the width of the inverter buffer which drives data onto the TSV network and the width of the transmission gate. Small transmission gate widths limit the amount of current that can pass through the gate even in its low impedance state, drastically reducing the shift frequency even at large driver widths. Likewise, a small driver width limits scan frequency even at large gate widths because it is

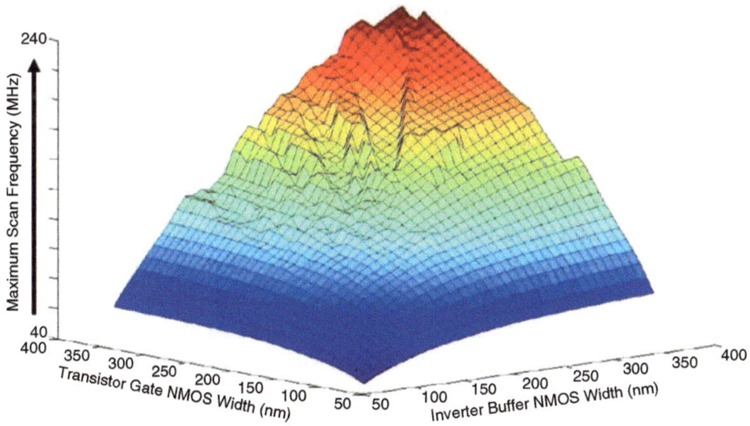

Fig. 5.12 Maximum scan-out frequency in a 11 TSV network with varying driver and transmission gate width for a 32 nm technology

incapable of sourcing or syncing enough current to quickly charge or discharge the TSV network capacitance.

As expected, at similar widths the 32 nm technology resulted in lower shift frequency when compared to the 45 nm technology, but both models showed that reasonable shift frequencies can be achieved without significant increases in the buffer or gate W/L ratios. For example, for an NMOS W/L ratio of 2/1, at 45 nm the maximum achievable shift frequency is 71 MHz, while at 32 nm the maximum shift frequency is 44 MHz. Increasing the ratio to 3/1 results in shift frequencies of 86 and 55 MHz, respectively, and at 4/1 the frequency becomes 98 and 62 MHz, respectively.

The size of the NMOS and PMOS transistors in the TSV driver can be fine-tuned to coincide with the scan frequency achievable from the automated test equipment. The maximum scan frequency of the pre-bond scan test increases significantly with driver size. To achieve a scan frequency of 200 MHz at 45 nm requires NMOS and PMOS width to length rations of about 5/1. After a point, the drivers are made too large and there is a significant drawback in the higher power consumption and parasitic capacitance of the drivers. However, Figs. 5.11 and 5.12 show that scan frequencies above 200 MHz are achievable without significantly larger drivers.

Next, the effect of scan configuration on test time is discussed. In Sect. 5.2.1, several possible scan configurations were described—one in which the scan I/Os for a scan chain are on the same TSV network (Configuration A), one in which they are on separate networks (Configuration B), and one in which multiple scan chains share the same TSV networks. While there are many possible ways that this third configuration can be constructed, this section will examine two examples, with the example from Fig. 5.5a being Configuration C and Fig. 5.5b being Configuration D. The scan frequency, scan chain length, number of scan chains, and number of

TSV networks determine which configuration results in a lower test time. Three examples are presented to highlight this issue, with test times determined as per Eqs. 5.1–5.4.

For Die 0 of the 4-layer FFT benchmark, if 50 scan chains are created, the result is a maximum scan chain length of 402 cells and 633 stuck-at test patterns. It is assumed that a probe card with 100 probe needles for contacting TSV networks is utilized. It is further assumed that Configurations A and C utilize the maximum scan-in (185 MHz) and scan-out (98 MHz) clock frequencies. Configurations A and C can use different scan-in and scan-out frequencies because these two shift operations are not performed in parallel. However, scan-in and scan-out are not overlapped. Configurations C and D utilize a broadcast scan-in, where patterns for each scan chain are combined to form a single, longer pattern. In this case, Configuration A requires 4.0 ms to complete the stuck-at scan test. Configuration B, operating only at 98 MHz, requires 2.6 ms because it can scan out test responses while scanning in the next test pattern. Configuration C requires 7.2 ms, because larger test patterns and the need to serially scan out two sets of test responses requires significantly more test time. Configuration D requires 3.9 ms, because broadcast scan-in requires additional test clock cycles that require more time than the simultaneous scan-out operations. In this example, Configurations A, B, C, and D require 50, 100, 25, and 75 TSV networks, respectively. Thus, if the probe card only supported 75 TSV networks instead of 100, then Configuration D would result in the shortest test time because Configuration B would require multiple touchdowns.

On the other hand, if the die has 100 scan chains, the maximum scan chain length is 202 cells and ATPG results in 640 stuck-at patterns. Because Configuration A can handle a maximum of 100 scan chains in a single touchdown, it needs to contact the die only once. This results in a test time of 2.0 ms. Configuration B requires two touchdowns; each time it is only capable of loading and unloading 50 scan chains. It is assumed that the die is partitioned into separate test modules each of 50 scan chains such that coverage remains high. In this case, Configuration B requires 4.6 ms for test plus the time required to align the probe card and for the second touchdown. Configuration C requires 3.7 ms and a single touchdown. Configuration D requires 6.6 ms and two touchdowns.

Our final example contains 150 scan chains, with a maximum scan chain length of 134 cells and 637 stuck-at patterns. Under these conditions, Configuration A requires two touchdowns, Configuration B requires three touchdowns, and Configuration D requires five touchdowns. Therefore, test times utilizing these configurations will be significantly larger than for Configuration C. Discounting the number of touchdowns required by Configurations A, B, and D, they have test times of 2.6, 2.7, and 6.5 ms, respectively. Configuration C needs only 1.6 ms for test, and because it requires only a single touchdown it will be the most cost-effective configuration for performing pre-bond scan test.

As stated earlier, not all TSVs need to be contacted for die logic testing. This is an important advantage, especially if TSV or microbump damage due to probing is a concern. Table 5.6 shows what percentage of TSVs must be contacted depending

Table 5.6 Percentage of TSVs that must be contacted for Die 0, as a function of the number of scan chains and scan configuration

Scan chain	% of TSVs to be contacted	
Number	Configuration A	Configuration B
25	2.7	5.3
50	5.3	10.7
75	8.0	16.0
100	10.7	21.4

on the number of scan chains present on the die and the scan configuration used. If oversize probe pads are used with the same number of scan chains, significant overhead must be incurred due to the large number of probe pads (even with test compression solutions). If the number of probe pads is limited, the test time will be higher because of constraints on the number of scan chains.

5.2.3 *Summary*

Section Summary

- Pre-bond scan test can be performed through TSV probing to enable rapid, high-bandwidth, low-area-overhead logic test.
- Various architecture configurations can be utilized to tailor scan test to maximally utilize the number of pre-bond TSV networks available for test.
- Scan shift frequency is determined by the architecture configuration utilized, the size of the TSV networks, and the transistor sizing of the GSFs.
- Power delivery through TSVs and probe needles is shown to be adequate for pre-bond scan test.

5.3 Conclusions

An extension of the TSV probing architecture of Chap. 4 has been introduced that can be used not only for pre-bond TSV test, but also for full-scan pre-bond die logic test. Scan chains are reconfigured into a pre-bond state to use TSV networks for scan I/O while preserving significant test parallelism and not requiring many oversized probe pads. HSPICE simulation results highlight the feasibility and effectiveness of this approach. Simulations show that the current needed for testing can be supplied through TSVs and probe needle tips. Test clock frequencies also remain relatively high even with the increased TSV network capacitance. The clock frequency can be tuned by adjusting the strength of drivers of the TSV network. The area overhead of this approach was estimated to be between 1.0% and 2.9% across the dies with TSVs for the two 4-die logic-on-logic 3D stacks presented in this chapter.

Chapter 6
Overcoming the Timing Overhead of Test Architectures on Inter-Die Critical Paths

6.1 Introduction

As discussed in previous chapters, 3D ICs require both pre-bond and post-bond testing to ensure stack yield. The goal of pre-bond testing is to ensure that only known good die (KGD) are bonded together to form a stack. Post-bond test ensures the functionality of the complete stack and screens for defects introduced in alignment and bonding. In order to enable both pre- and post-bond testing, die level wrappers have been proposed in the literature [35, 93] and are discussed further in Chap. 7. These die wrappers include boundary scan cells at the interface between die logic and TSVs to add controllability and observability at the TSV. Chapters 4 and 5 described how modified boundary scan cells, called gated scan flops (GSFs), can be used along with pre-bond probing for pre-bond TSV and structural logic tests. Together, published work such as [35, 76, 93, 94] shows that die wrappers can not only be used as a standard post-bond test interface, but also for pre-bond KGD tests.

Figure 6.1 gives an example of a two-die logic-on-logic stack, where TSVs are utilized as fast interconnects between logic on two separate dies. One of the drawbacks of inserting die-boundary cells is the adverse timing impact on these TSV functional paths. A reduction in latency using short TSV interconnects is one of the key drivers for 3D technology in both logic-on-logic and memory-on-logic applications. Adding boundary flops to TSVs between two layers adds two additional clocked stages to a functional path, which would otherwise not exist in a two-dimensional (2D) design. Bypass paths can be added to the boundary scan cells to multiplex the functional input between being latched in the flop or being output directly to/from the TSV. During functional mode, the bypass path is active and a signal traveling across dies is never latched in the boundary register; however, the bypass path still introduces additional path latency. It is important to note that additional latency is added by test architectures not just to logic-on-logic stacks, but also to memory-on-logic and memory-on-memory stacks that contain dies with die wrapper boundary registers.

B. Noia and K. Chakrabarty, *Design-for-Test and Test Optimization Techniques for TSV-based 3D Stacked ICs*, DOI 10.1007/978-3-319-02378-6_6,

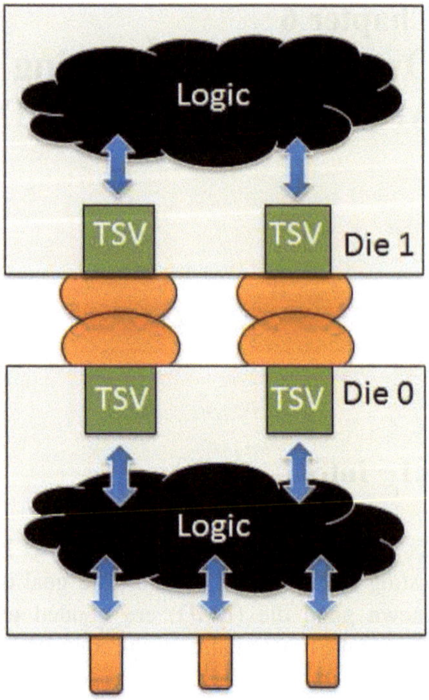

Fig. 6.1 Example of a two-die logic-on-logic stack

Retiming is an algorithmic approach to improving multiple aspects of circuit design post-synthesis by moving the positions of registers with respect to combinational logic while preserving circuit functionality [95]. Retiming methods have been developed to target a number of circuit features, including minimizing clock period [96], reducing power consumption [97], and improving testability by reducing feedback dependency [98]. Previous literature focuses on retiming in 2D circuits after synthesis but before test insertion. This is because before synthesis, the RTL behavioral circuit model does not contain register locations.

This chapter extends the concept of retiming to utilize retiming algorithms to recover the latency added by boundary scan cell bypass paths in a 3D IC. Retiming in this case is performed after synthesis and 3D boundary cell insertion. Two-dimensional retiming methods can be reused after test-insertion by fixing the location of die wrapper boundary flops so that they remain at the logic/TSV interface during retiming. This requirement ensures that die logic is not moved across dies when retiming is performed on a complete stack. A bypass path is added to each register so that in functional mode, data does not need to be latched, thereby replacing extra clock stages with added latency. Retiming is then performed to recover the added latency of the bypass along the TSV path. An additional step is added to standard retiming methods, whereby complex logic gates that prevent register relocation due to latency violations are decomposed into their basic logic cells from the circuit library. Furthermore, a logic redistribution algorithm, whereby certain logic cells on a critical path can be moved from one die to its adjacent die, is applied to die-level retiming in order to provide better delay reduction results. This step can allow for a redistribution of circuit delay to non-critical paths from low-slack paths during retiming.

6.1.1 The Impact of Die Wrappers on Functional Latency

Die wrappers are discussed in Chap. 7 as a means for standardizing the test interface at die level for pre-bond and post-bond test. Die wrappers leverage many of the features of the IEEE 1500 standard wrapper for embedded cores [20]. A wrapper instruction register (WIR) is loaded with instructions to switch the wrapper between functional mode and test states. Switch boxes multiplex between a post-bond test mode and a reduced pin-count pre-bond test mode. A wrapper boundary register (WBR) consists of scan flops at the die interface between the next die in the stack (or the primary inputs and outputs in the case of the lowest die in the stack) and the die's internal logic.

The WBR provides the means to enhance pre-bond test coverage. During pre-bond scan test, die boundary registers provide controllability and observability at each TSV. Without boundary registers, untestable logic exists between die-internal scan chains and TSVs. In a 3D benchmark circuit without boundary registers, stuck-at and transition fault coverage for a die was shown to be as low as 45% [94]. With the addition of die boundary registers on the same die, the fault coverage for the die was shown to be above 99%.

Although boundary registers, with or without GSFs, are necessary for pre-bond fault coverage, prior work has not addressed the latency overhead that boundary registers contribute to the signal paths that cross die boundaries. One of the advantages of 3D stacked ICs is the reduction in interconnect delay gained by utilizing short TSVs in place of long 2D interconnects. For example, in a memory-on-logic stack, latency on the TSV path directly impacts memory access time. In logic-on-logic stacks, least-slack paths are split between die layers to increase operating frequency [44,45].

In 3D circuits, especially logic-on-logic stacks, the latency improvements of 3D integration are most relevant when critical paths are split between two or more dies. The use of low-latency TSVs instead of long 2D interconnects increases the slack of critical paths and therefore allows for a faster functional clock. Unlike wrapping 2D test modules, for example as per the 1500 standard, if die-level WBRs are added to a 3D circuit, they must be unavoidably on critical paths. This is because they will exist on every path that utilizes a TSV or has to interface with a TSV on another die. This removes slack from the same critical paths that are split between dies. Through retiming, slack can be redistributed throughout a circuit in such a way that cross-die critical paths can meet their pre-wrapper-insertion timing requirements.

In this chapter, it is assumed that a bypass path is added to the GSF boundary registers of the die wrapper, as shown in Fig. 6.2. In functional mode, the bypass signal is asserted to route the functional input directly to the TSV, circumventing

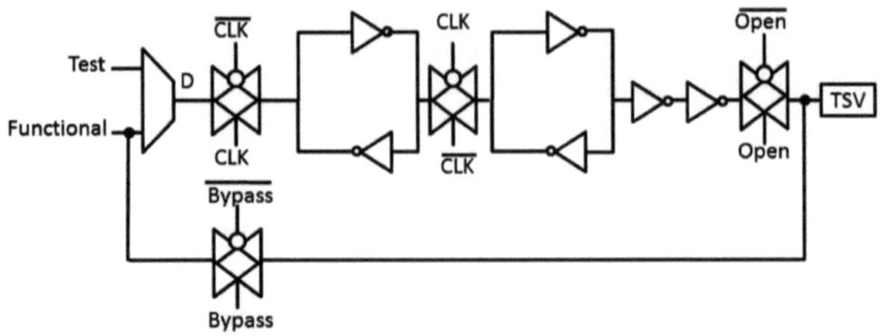

Fig. 6.2 An example gate-level design of a gated scan flop with a bypass path

the need to latch functional data in the boundary registers. However, even when using a bypass path, the addition of boundary registers still increases latency on TSV paths. This chapter introduces methods to recover this latency through register retiming.

6.1.2 Register Retiming and Its Applicability to Delay Recovery

Register retiming is a post-synthesis, algorithmic approach to circuit optimization first introduced in [95]. While retiming can be used to target a number of applications, including reducing power consumption [97] or enhancing testability [98], most retiming algorithms focus on clock-period reduction. During retiming, slack is moved from paths with excess slack to least-slack paths without altering the function of the circuit by moving the location of registers in relation to combinational logic. Transforms are defined to limit how and where registers can be moved by retiming algorithms to preserve circuit functionality [96].

In the simplest algorithms, retiming is performed by representing a circuit as a directed graph where vertices represent logic gates and edges represent connections between logic elements [95]. The weight of each edge represents the number of registers present between two logic elements. The propagation delay through each vertex is calculated as a function of the number of logic elements through a directed path that does not contain a register. During retiming, registers are moved between graph edges to reduce the delay values for graph paths, a step that leads to a reduction in the clock period.

While most retiming algorithms are used after synthesis, accurately approximating path delay from a structural circuit definition is difficult without placement. Furthermore, the movement of registers can alter interconnect lengths. These changes are difficult to account for with only a structural description of a circuit. The authors of [99] carry out retiming after performing an initial placement in order to model interconnect delay more accurately. During retiming, it is assumed that combinational logic placement remains unchanged but that a retimed register would be placed at a range around the geometric center of its fan-in and fan-out cones based on slack.

Retiming has been utilized in the literature to recover the delay added to a path where a standard flip-flop is converted into a scan flip-flop [100]. The extra delay is caused by the addition of a multiplexer to select between a functional and test input. Retiming is then performed to move the flop portion of the scan flop in relation to the added multiplexer. This method is not applicable for die-boundary registers, however, as these registers cannot be moved. Furthermore, in functional mode, the wrapper boundary registers are not used to latch data and are instead set to a bypass mode that is independent of the multiplexer required to select between functional and test inputs.

In this chapter, retiming is utilized after synthesis and test-architecture insertion in a 3D stack to recover the additional delay added to a TSV path by boundary registers. In order to present simulations results, the retiming algorithm of Synopsys Design Compiler is used, which performs retiming first to minimize clock period and then to minimize register count. This is followed by a combinational logic optimization step in view of changes to loading on cells and critical paths that may have occurred after the movement of registers. An additional step of logic decomposition is added to the retiming procedure—if a complex logic gate is preventing further retiming on a critical path that does not satisfy slack constraints, then the gate is decomposed into simple gates from the cell library that together perform the equivalent Boolean function. Retiming is then performed again to determine if logical decomposition further reduces latency. Furthermore, logic can be shifted between dies to achieve better retiming results.

The rest of this chapter is organized as follows. Section 6.2 provides a motivating example for post-test-insertion retiming and outlines the retiming methods used in this chapter. Section 6.2.1 provides a detailed description of the test insertion and retiming flow used to recover delay overhead. Section 6.2.2 discusses the design flow and algorithm used to redistribute logic during die-level retiming. Section 6.2.3 presents retiming results for two-, three-, and four-die stacks with TSVs in logic-on-logic 3D benchmarks as well as a modular processor benchmark. Finally, Sect. 6.3 concludes the chapter.

6.2 Post-DfT-Insertion Retiming in 3D Stacked Circuits

Figure 6.3 illustrates retiming performed on an example circuit on the bottom die of a 3D stack. Figure 6.3a shows the circuit prior to wrapper insertion and retiming. Inputs A, B, C, and D are primary inputs. The circuit contains four flip flops (FF), labeled f_1 through f_4. Several logic gates exist between the flops and the TSV that connect to the next die in the stack. These are part of an incomplete logic circuit without the connected logic in the other stack tiers and are not observable during pre-bond scan test. Basic and complex logic cells are annotated with a delay value representing delay through the cell as a fraction of the clock cycle. For example, a value of 0.5 means that rise and fall delays from the inputs to the output of a logic gate is half of one clock period. This example, and the timing data

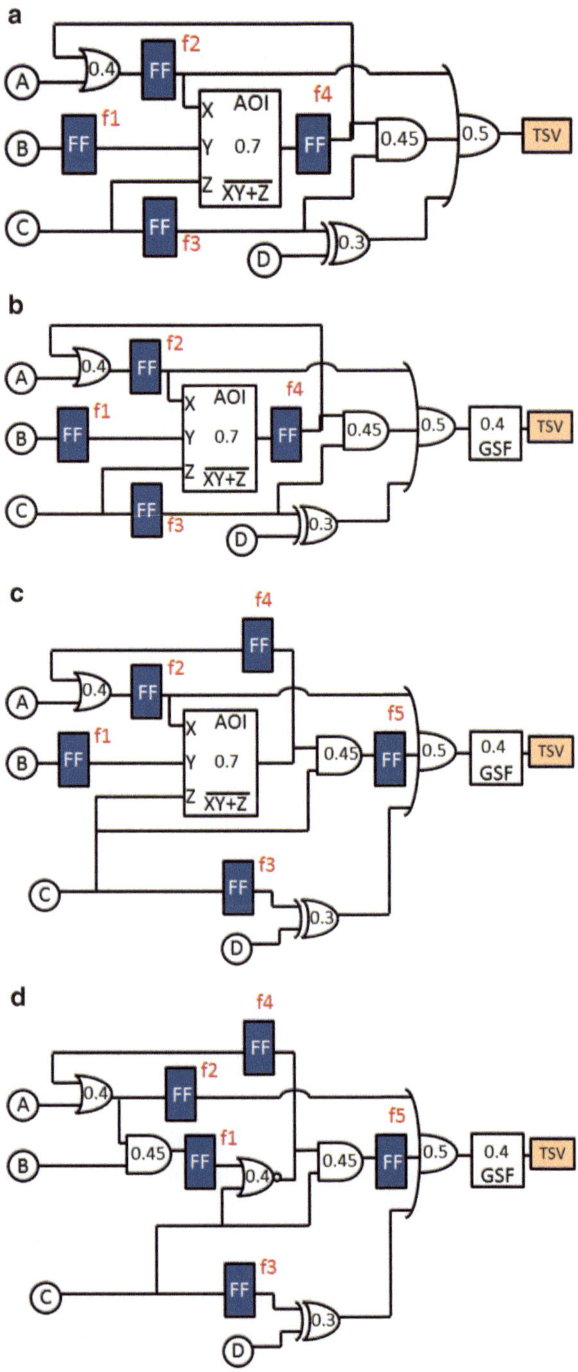

Fig. 6.3 Boundary register insertion and retiming (including logic decomposition): (**a**) design before register insertion, (**b**) design after register insertion, (**c**) first register movement, (**d**) logic decomposition and second register movement

presented, is simplified for demonstration and does not consider additional data such as interconnect delay, fan-out, clock skew, or flop setup and hold times. It is assumed for this example that no path from a primary input or flop to another flop or to the TSV can have a delay greater than one clock period. In Fig. 6.3a, the longest delay is 0.95 of a clock period from flop f_4 to the TSV, so this timing constraint is satisfied.

In order to provide controllability and observability, and to present a standard test interface, a wrapper is next added to the die. For the example circuit, a boundary GSF is inserted between the die logic and the TSV as shown in Fig. 6.3b. In its bypass mode, the GSF contributes an additional 0.4 delay to any path through the TSV. For the path from flop f_2 to the TSV, this is acceptable, because the overall path delay is only 0.9 clock cycles. However, three paths now violate the timing constraints—the paths from f_3 and f_4 to the TSV at a delay of 1.35 clock cycles each, and the path from D to the TSV at a delay of 1.2 clock cycles.

Retiming is performed to recover the latency added to paths through the TSV by moving registers to ensure that no path violates the timing constraint. In the first retiming step, flops f_3 and f_4 on the AND gate inputs are pushed forward to its output as shown in Fig. 6.3c. The AND gate's inputs now come directly from C and the complex AOI gate without latching, and its output is instead latched. This requires the addition of an extra flop f_5 at the AND gate's output. The extra flop is created because a flop (f_4) is still required on the feedback loop from the AOI gate to flop f_2 in order to preserve functionality. Likewise, a flop (f_3) is still required between C and the XOR gate. Now the path from flop f_5 to the TSV satisfies the timing constraint with a delay of 0.9 cycles. On the other hand, the path from flop f_2 to flop f_5 now violates the constraint with a delay of 1.15 clock cycles.

At this point, further retiming cannot satisfy the timing constraint. If f_2 is pushed to the output of the AOI gate, then the path from A to f_2 will violate timing constraints. If f_5 is moved to the inputs of the AND gate, then the circuit returns to its state in Fig. 6.3b. The AOI gate does not allow any further retiming, so in an effort to provide greater flexibility to the retiming algorithm, it is decomposed into basic logic gates from the cell library—an AND and NOR gate. Compared to its decomposed gates, the AOI gate provided for a shorter delay for the Boolean function it performed and required less die area. However, the basic logic gates allow more avenues for retiming. Flops f_1 and f_2 can now be pushed from the inputs of the to the output of the AND gate added after decomposition, as shown in Fig. 6.3d. All paths now satisfy the timing constraint and retiming is complete. Although there is less slack in the circuit overall after the addition of the boundary GSF and the decomposition of the AOI gate, the slack is redistributed among all paths so that no single path experiences a timing violation.

It is important to note that the delay values given for GSFs and gates in the motivational example is only meant to illustrate why and how retiming may be utilized. These do not present realistic delay values. In simulation, the delay of the GSF in bypass mode that is added to each die on an inter-die path amounted to 3–4 inverters designed in the same technology node. Therefore, a path that spans one bonded interface between dies has an additional delay equivalent to seven inverters.

If a path crosses two bonded interfaces, the additional delay is equivalent to 10–11 inverters, and so forth. The additional delay caused by the GSFs in any given design will depend on the design of the GSF and the manufacturing technology used.

The rest of this chapter refers to a metric designated as "% delay recovered" to denote the quality of retiming results. This does not refer to the percentage of recovered paths. The delay recovered is calculated from the actual operational frequency of the circuit. Three minimum timing values are produced for a given design. The first (*original*) value is the frequency of the circuit before DfT insertion. The second (*inserted*) value is the frequency of the circuit after DfT insertion, which in all examined benchmarks was less than the original frequency due to slack violations. The third (*retimed*) value is the frequency of the circuit after retiming, after which slack is redistributed in an attempt to allow the circuit to run at the original frequency value. From these variables, the equations

$$a = inserted - original$$

$$b = retimed - original$$

are defined and "% delay recovered" is calculated as

$$\frac{a-b}{a} \cdot 100.$$

When *inserted* equals *retimed*, then 0% of the delay is recovered. When *retimed* equals *original*, then 100% delay is recovered.

6.2.1 Method for Die- and Stack-Level Retiming

Retiming can be performed either at the die- or stack-level. In the example of Fig. 6.3, retiming was performed at the die-level, or in other words, on a single die of a 3D stack without knowledge of the circuitry on other dies. Die-level retiming allows greater control over the redistribution of slack throughout the stack. For example, a designer may not want to move any registers on a particular die D but would still like to recover the additional latency of adding a wrapper to that die. In this case, additional (dummy) delay can be added to the TSV paths on the dies adjacent to D. Retiming can then be performed on the adjacent dies in an attempt to recover the additional path delay due to the wrapper cells in the adjacent dies and the wrapper cells of D.

A flowchart of the steps required for post-wrapper insertion die-level retiming is shown in Fig. 6.4. First, the design is synthesized into a structural circuit definition. In die-level retiming, paths crossing the TSV interface are incomplete and the total delay across these paths cannot be considered. Because paths crossing die boundaries are likely to be the least-slack paths in a 3D stack [44, 45], the clock period for the stack may be too large to provide a tight timing constraint when

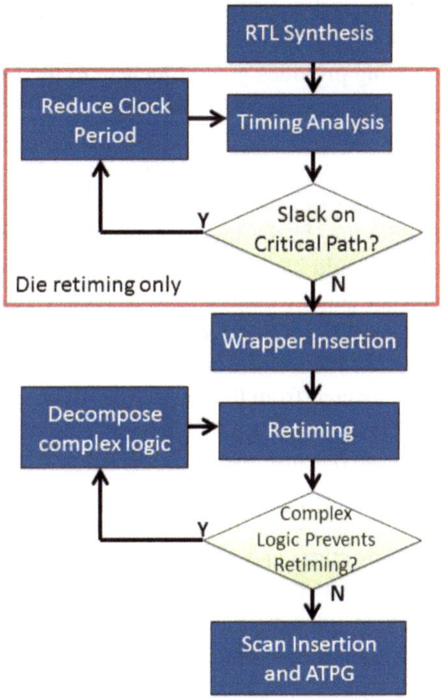

Fig. 6.4 Flowchart for retiming of a 3D stack at either die- or stack-level

considering a single die. In order to determine an appropriate timing target, timing analysis is performed to identify the amount of slack on the least-slack path of the die. The target clock period for retiming is incrementally reduced until the least-slack path has no positive slack. Wrapper insertion is then performed, adding delay to the TSV paths equal to the bypass path through a boundary GSF. During retiming, boundary GSFs are fixed so that the retiming algorithm does not consider them as movable registers nor does it attempt to move logic or other registers past the GSFs. Timing information for the logic gates, flip flops, and GSFs are extracted from the cell library. After the retiming algorithm has executed, timing analysis is again performed to determine if all paths in the die satisfy the target timing constraint. If they do not, the path that has the most negative slack is examined to determine if complex logic gates on the path may be preventing retiming. If complex logic gates are preventing retiming, they are decomposed into simple logic cells from the cell library and retiming is performed again. This process continues until all paths meet the timing target or no complex logic gates are restricting retiming. Finally, scan insertion and ATPG are performed for the die.

While die-level retiming does not consider the total path delay for paths that cross die boundaries, stack-level retiming can exploit this added degree of freedom. In stack-level retiming, the complete stack is retimed as a monolithic entity. The boundary GSFs are once again fixed during retiming to prevent the movement of logic from one die to another. During stack retiming, the intended clock frequency for the stack can be used as a timing target because all circuit paths are known. While die-level retiming provides more control over slack redistribution, stack-level retiming provides greater leeway to the retiming algorithm. To illustrate this point, consider a three-die stack where each boundary GSF adds an additional 0.2 clock period delay to a path. A path crossing all three dies from the bottom to the top of the stack would experience an additional 0.8 clock period delay after wrapper insertion—0.2 delay from both the bottom and top dies and 0.4 delay from the middle die. In die-level retiming of all the dies in the stack, 0.2 clock periods of delay would have to be recovered in each of the top and bottom dies of the stack, and 0.4 delay would have to be recovered in the middle die. In stack-level retiming, extra slack can be redistributed to the path regardless of which die can supply the slack. To recover the entire 0.8 clock period of delay added to the path, 0.1 delay can be recovered from the lowest die, 0.3 from the middle die, and 0.4 from the top die.

The flow for stack-level retiming is similar to the die-level retiming flow of Fig. 6.4. Because the clock period of the stack is known, it can be used as a timing constraint for retiming. For this reason, no timing analysis or tightening of the clock period must be performed before wrapper insertion. Retiming, logic decomposition, scan insertion, and ATPG are performed as they were for die-level retiming.

The drawback of die-level retiming—namely that a die-level retiming may result in inter-die paths that continue to violate timing constraints after wrapper insertion when a stack-level retiming solution would satisfy timing constraints—motivates the need for an enhancement to die-level retiming. In this chapter, a logic redistribution algorithm, which is described in detail in Sect. 6.2.2, is introduced to better make use of excess slack during die-level retiming. Consider, for example, the inter-die path shown in Fig. 6.5 between flip-flop A (FF_A) on Die 0 and flip-flop B (FF_B) on Die 1 in a hypothetical stack. Figure 6.5a shows the example path before wrapper insertion or retiming. The full path consists of ten logic cells, each with a delay of 1 ns, and two TSVs, each also with a delay of 1 ns. The entire path has a delay of 12 ns, and no path in the circuit can have a delay greater than 13 ns. Each die contains five logic cells and one TSV of the path. There exists enough slack on Die 0 such that FF_A can be moved toward its TSV by one logic cell, and on Die 1 there is enough slack for FF_B to be moved toward its TSV by four logic cells. Figure 6.5b shows the path after wrapper cell insertion, where each wrapper cell (GSF) contributes 3 ns of delay in functional mode. Now, the full path requires 18 ns for signal propagation, well above our limit of 13 ns. During die level retiming of Die 0, as shown in Fig. 6.5c, FF_A can only be moved to regain 1 ns of delay, thus violating its timing constraints. For Die 1, FF_B will be moved to regain 3 ns of delay, recovering all of the additional delay. Thus, with die-level retiming, the path violates the overall timing constraint by 1 ns, even though there is additional slack on Die 1 that would be appropriately allocated during stack-level retiming. A die-level solution to this problem would be to move one logic cell, the grayed cell, from Die 0 to Die 1 by pushing it across the die boundary as shown in Fig. 6.5d. Now, when die-level retiming is performed, Die 1 will recover 4 ns of delay as in Fig. 6.5d and the timing constraint for the path will be met.

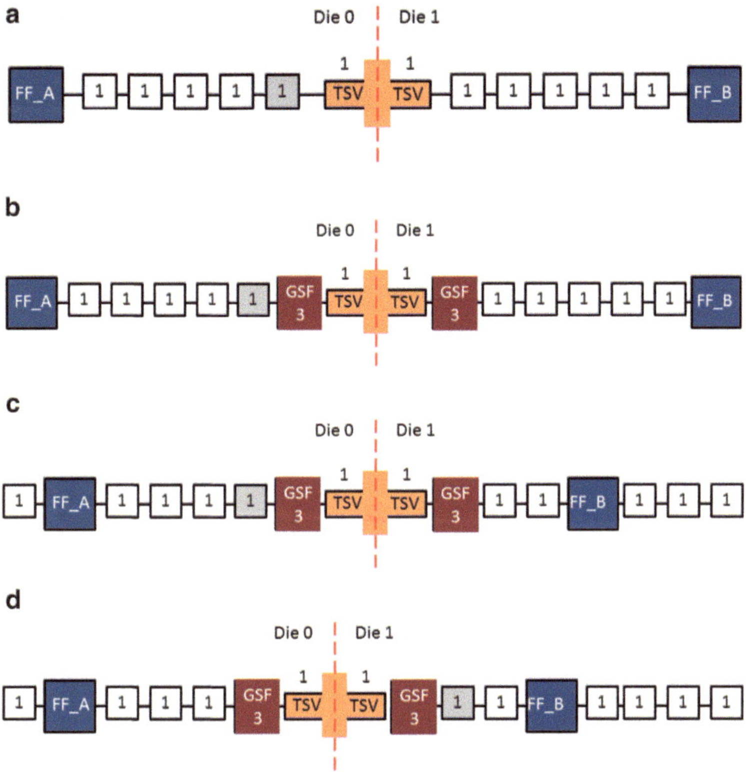

Fig. 6.5 Logic redistribution example showing: (**a**) design before register insertion, (**b**) design after register insertion, (**c**) die-level retiming before logic redistribution, (**d**) die-level retiming after logic redistribution

In all of the benchmarks presented in this chapter, the circuits have been partitioned for timing optimization. Before DfT insertion, the most critical path in each benchmark is an internal path in a die, since paths that would be critical in the stacked circuits utilize TSVs where available and are no longer critical paths. After DfT insertion, at least one inter-die path becomes the most critical path of the design and causes timing violations that were not present in the pre-test-insertion design. The retiming method described and utilized in this chapter does not attempt delay recovery on every path. It is an iterative process where the least-slack path that violates the timing constraint is first retimed, and then the next, and so forth until the timing constraint is no longer violated or the current least-slack path cannot be made to satisfy the timing constraint. Paths that do not violate the timing constraint after DfT insertion are not affected except in certain cases where they may be altered to retime a critical path.

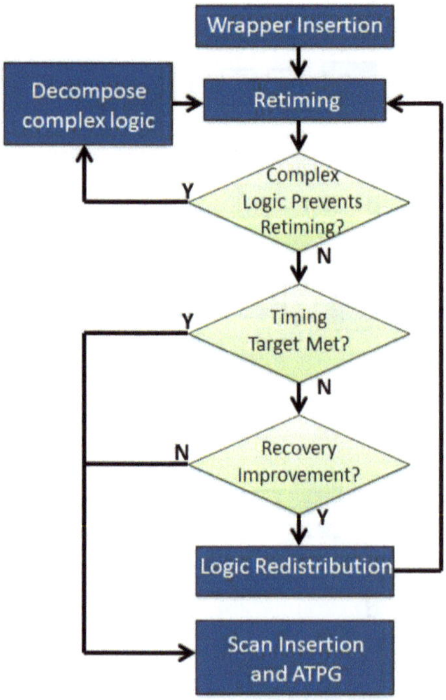

Fig. 6.6 Flowchart for die-level retiming with logic redistribution

6.2.2 *Algorithm for Logic Redistribution*

Figure 6.6 shows the insertion of the logic redistribution algorithm into the die-level retiming flow. A logic redistribution algorithm is inserted after complex logic decomposition. If timing targets are met, then no logic redistribution is necessary and scan insertion and ATPG can take place. If, on the other hand, timing targets are not met after die-level retiming, then logic redistribution can be performed in an attempt to achieve additional delay recovery. Logic redistribution is performed, followed by retiming and logic decomposition if necessary, until either no recovery improvement is achieved, in which case scan insertion and ATPG are performed on the last netlist that resulted in delay recovery improvement, or until all timing constraints are met.

For any given iteration, the logic redistribution algorithm attempts to shift logic one cell at a time on the most critical inter-die path from one die to its adjacent die. Retiming is then performed again, and if necessary another iteration of the logic redistribution algorithm takes place. This allows the algorithm to target either the same path for further improvement or a new path if that path becomes the predominant critical path.

The pseudocode for the logic redistribution algorithm (LogicRedis) is given in Algorithm 2. The algorithm requires as input the netlists for two dies, $DieA$ and $DieB$, that are adjacent to another and share the critical path $CritPathA$. The library of standard cells, $cellLibrary$, is also needed. The algorithm attempts to move one logic cell from the critical path on $DieA$ to $DieB$.

Algorithm 2 LogicRedis($DieA$, $DieB$, $CritPathA$, $cellLibrary$)

```
cellToMove = getCell(DieA, CritPathA);
fanIn = getFanIn(DieA, cellToMove, cellLibrary);
controlSig = getControlSignal(cellToMove, cellLibrary);
All variables are initialized at this point.
if fanIn OR controlSig then
   return {DieA, DieB};
else
   {DieA, DieB} = moveLogic(DieA, DieB, CritPathA);
end if
return {DieA, DieB};
```

The LogicRedis begins by identifying the logic cell that must be moved through the **getCell** function. This function takes the netlist for $DieA$ and the critical path identification and finds the logic cell on the critical path on $DieA$ closest to the TSV and returns its unique name and standard cell type, which are assigned to the data structure $cellToMove$. It is then necessary to determine whether or not the cell can be moved to $DieB$.

Logic redistribution is limited by the fact that TSVs between the adjacent dies cannot be added to or subtracted from the stack design. Whether a logic cell is being moved from a transmitting die to a receiving die or vice versa, a logic cell with a fan-in greater than one or that requires a control signal (such as a transmission gate) cannot be moved without adding or removing TSVs. There are two exceptions to this rule. The first is a logic cell with fan-in greater than two but with all inputs except for the TSV input tied to either V_{dd} or ground. The second is a logic cell with a control signal that is tied to V_{dd} or ground. The **getFanIn** function returns true if the logic cell ($cellToMove$) passed to it has a fan-in greater than one and at least two fan-in nets are not connected to V_{dd} or ground. The **getControlSignal** function returns true if the passed logic cell requires a control signal and that signal is not tied to V_{dd} or ground.

If moving the logic cell would violate our constraint that TSVs cannot be added or removed—that is, if either $fanIn$ or $controlSig$ is true—then the input netlists $DieA$ and $DieB$ are returned without modification. If, instead, the logic cell can be safely moved from $DieA$ to $DieB$, then the **moveLogic** function performs the logic redistribution and returns new netlists for $DieA$ and $DieB$. These new netlists are then returned in order to perform die-level retiming once again such that the new netlists may lead to better delay recovery.

The **LogicRedis** algorithm terminates under two circumstances. The first is after moving a single logic cell, in which case both $fanIn$ and $controlSig$ are false and the $DieA$ and $DieB$ netlists are updated with the moved logic and returned. If the algorithm is terminated in this way, then retiming is performed again and the result is checked to see if the timing target is met or if there is recovery improvement as per Fig. 6.6. The second termination criterion is that no logic cell can be moved without adding or subtracting a TSV, in which case either or both of the $fanIn$ and $controlSig$ variables will be true. In this case, the $DieA$ and $DieB$ netlists are returned unchanged and no further retiming is possible.

The algorithm for logic movement is of complexity $\mathcal{O}(n)$ with regard to the size of the netlist of $DieA$. The **getCell** function is $\mathcal{O}(n)$ as it must search the netlist for the cell to be retimed, whose inputs and outputs match those of the critical path identification. The functions **getFanIn** and **getControlSignal** are constant-time, as they only require checking the inputs and outputs of a standard cell definition. Likewise, the **moveLogic** function is constant time as it adding or removing the previously identified logic cell and two wire definitions from each netlist.

The limitation on adding or removing TSVs during logic redistribution limits how close die-level retiming can come to stack-level retiming. Consider a worst-case retiming scenario in a two-die stack where a cross-die critical path exists such that most of the path logic is on Die 0 while only a single logic cell of the path is on Die 1. During stack level retiming, the burden of recovering the timing latency of the path is almost entirely on Die 0, while the retiming of Die 1 may not consider the path to be critical at all if only a small portion of the path is on Die 1. In this case, it is beneficial to move some of the path logic from Die 0 to Die 1 to evenly distribute the retiming burden between the dies, or to move all of that path logic to Die 1 if Die 1 has a significant amount of slack that can be moved onto the path while Die 0 has none. If the entire path across both dies consists of logic cells that satisfy our TSV constraints—that is, if any or all cells can be moved between the dies—then die-level retiming can achieve the same results as stack-level retiming for that particular path. This is because a logic distribution can be found such that each die is capable of providing all the extra slack it can to that path. In real circuits, however, it is likely that only some logic on some critical paths will meet the TSV requirements for logic redistribution, and therefore stack-level retiming will outperform die-level retiming.

To implement the retiming flows in practice, the retiming tool utilized in this chapter (Synopsys Design Compiler) reports which gate on the least-slack path is preventing further retiming when a timing target is not met. If the timing target is not met and a gate is identified, the decomposition algorithm implemented in this paper checks to see if the gate is a complex gate. If so, the gate is decomposed and retiming is attempted again. If no gate is flagged and the timing target is met, then 100% delay recovery is achieved and the retiming algorithm terminates. If the timing target is not met, but there was a recovery improvement over the previous retiming result, then logic redistribution is attempted. Only in the case where logic decomposition is not necessary and logic redistribution has been performed with no delay recovery improvement over the previous retiming attempt does the retiming algorithm terminate without 100% delay recovery.

6.2.3 The Effectiveness of Retiming in Recovering Test-Architecture-Induced Delay

In this section, the effectiveness and impact of post-DfT-insertion retiming is examined in the context of several benchmark circuits that have been partitioned into two-, three-, and four-die stacks. Two benchmarks are used—the performance-optimized data encryption standard (DES) circuit and the Fast Fourier Transform (FFT) circuit from the IWLS 2005 OpenCore benchmarks [85]. The DES circuit contains 26,000 gates and 2,000 flops, while the FFT circuit contains 299,273 gates with 19,962 flops. They contain no embedded, wrapped modules or black boxes, so retiming can target any gates or flops in the design. The DES circuit was partitioned into two-, three-, and four-die stacks and the FFT circuit was partitioned into two- and four-die stacks using the Nangate open cell library [101] and a placement engine that optimized timing.

A third OpenCore benchmark—the OpenRISC1200 (OR1200) 32-bit scalar RISC processor—is used as a benchmark with modules that cannot be retimed. The OR1200 has a five stage integer pipeline and IEEE 754-compliant single precision floating point unit. The OR1200 implementation used in this chapter contains a one-way direct-mapped 8 KB data cache and one-way direct-mapped 8 KB instruction cache. The processor contains 15,000 cells with 1,850 flip-flops and utilizes a 250 MHz clock frequency.

No 3D modular benchmarks yet exist in the open literature, so a benchmark for this chapter was created by partitioning the OR1200 processor across two dies. Many modules were fixed, meaning that they could not be retimed, and were placed in their entirety on one die or the other. Fixed modules generally include modules where retiming may move critical registers that can interfere with timing closure, such as the decode, execute, memory, and writeback stages of the processor pipeline. The modules that were allowed to be retimed are the debug unit, exception logic, floating point unit, freeze logic, instruction fetch, load/store unit, programmable interrupt controller, power management, and SPR interface. These modules were split across the dies in such a way that logic that may be retimed existed on both sides of each TSV. This was done because, barring interconnects, no logic existed between modules. The modules that may be retimed in this benchmark take the place of complex logic between modules. If a fixed module were on either side or both sides of a TSV belonging to a critical path, then retiming may have no impact on delay recovery.

To approximate the bypass mode delay of a GSF, a GSF was created in HSPICE using a low-power 45 nm process [56]. Input-to-output rise and fall times were determined using simulation. To model wrapper insertion, a number of inverters from the cell library were added before each TSV and each face pad for TSV bonding to approximate the delay of bypass mode GSFs. All of the data given in this section utilizes the inverter delay approximation for the GSF. In all benchmarks presented, the delay of the GSF was significant enough to cause slack violations after DfT insertion.

Table 6.1 shows retiming results for the (a) two-, (b) three-, and (c) four-die DES stacks. The columns present data first for die-level retiming, starting with the lowest die in the stack (Die 0) and moving to the highest die in the stack. The last column shows data for stack-level retiming. The first row of the table lists the percent of delay recovered during retiming on the least-slack TSV path. A value of 100 means that all delay on all TSV paths was recovered. The second row indicates the area overhead of DfT insertion as a percentage of total area, or the sum of cell area and

Table 6.1 A comparison of delay, area, and pattern count results for die- and stack-level retiming for the DES logic circuit partitioned across (a) two, (b) three, and (c) four dies

(a) Two-die stack

	Die 0	Die 1	Complete stack
% Delay recovered	100	100	100
% Area overhead post-wrapper insertion	16.3	16.4	18.7
% Area overhead post-retiming	12.4	13.4	16.6
% change in pattern count	−7.4	3.0	10.9

(b) Three-die stack

	Die 0	Die 1	Die 2	Complete stack
% Delay recovered	100	100	100	100
% Area overhead post-wrapper insertion	20.0	29.8	26.2	26.2
% Area overhead post-retiming	19.7	29.1	24.2	25.0
% change in pattern count	3.3	6.2	−1.4	12.7

(c) Four-die stack

	Die 0	Die 1	Die 2	Die 3	Complete stack
% Delay recovered	100	100	60	100	100
% Area overhead post-wrapper insertion	22.7	35.5	35.6	28.5	31.9
% Area overhead post-retiming	16.1	34.1	34.6	25.9	27.5
% change in pattern count	−2.5	−4.2	0.8	5.3	8.1

interconnect area. Interconnect area is estimated from wire load models included with the cell library. The third row indicates the percentage of total area overhead after retiming. The last row presents the percentage change in pattern count for stuck-at patterns between ATPG before DfT-insertion and ATPG after retiming. A negative value indicates that pattern count decreased. Table 6.2 shows similar results for the FFT two- and four-die benchmarks.

Table 6.3 shows the change in pattern count and change in the statistical delay quality level (SDQL) reported by a commercial tool for small-delay defects. Table 6.3a provides results for the two-, three-, and four-die DES stacks and Table 6.3b provides results for the two- and four-die FFT stacks. Each row shows results for a stack of different size. The first column shows the SDQL after boundary

Table 6.2 A comparison of delay, area, and pattern count results for die- and stack-level retiming for the FFT logic circuit partitioned across (a) two and (b) four dies

(a) Two-die stack

	Die 0	Die 1	Complete stack
% Delay recovered	100	100	100
% Area overhead post-wrapper insertion	0.9	1.1	1.2
% Area overhead post-retiming	1.1	1.1	1.2
% change in pattern count	3.9	−3.0	6.4

(b) Four-die stack

	Die 0	Die 1	Die 2	Die 3	Complete stack
% Delay recovered	100	100	100	100	100
% Area overhead post-wrapper insertion	2.1	2.3	2.5	1.9	2.4
% Area overhead post-retiming	2.0	2.1	2.4	1.5	2.2
% change in pattern count	1.1	5.1	0.8	−0.4	7.8

Table 6.3 Change in SDQL and pattern count before and after retiming for the (a) two-, three-, and four-die DES stack and (b) two- and four-die FFT stack

(a) DES stack

# of dies in stack	Pre-retiming SDQL	Post-retiming SDQL	% change in SDQL	% change in pattern count
2-Die	183	172	−6.0	−2.1
3-Die	182	185	1.6	0.7
4-Die	178	182	2.2	3.0

(b) FFT stack

# of dies in stack	Pre-retiming SDQL	Post-retiming SDQL	% change in SDQL	% change in pattern count
2-Die	52385	51861	−1.0	0.9
4-Die	51977	52741	2.5	2.8

cell insertion and before retiming. Column II shows the SDQL after retiming, with column III providing a percent-change in SDQL between the pre- and post-retiming values. Column IV provides the overall percent-change in pattern count between pre- and post-retiming ATPG.

Table 6.4 provides retiming results for the four-die FFT benchmark when one or two dies are fixed, meaning that they cannot be retimed. Because there is added

Table 6.4 A comparison of delay recovery with fixed dies for the four-die FFT logic circuit partitioned considering (a) one and (b) two dies cannot be retimed

(a) One fixed die

	Delay recovery when retimed for:				
Fixed die	Die 0	Die 1	Die 2	Die 3	Complete stack
Die 0	–	100	100	100	100
Die 1	87.5	–	100	100	100
Die 2	100	100	–	100	100
Die 3	100	100	100	–	100

(b) Two fixed dies

	Delay recovery when retimed for:				
Fixed dies	Die 0	Die 1	Die 2	Die 3	Complete stack
Dies 0, 1	–	–	75.0	82.5	100
Dies 2, 3	50.0	62.5	–	–	71.9
Dies 0, 2	–	62.5	–	82.5	84.4
Dies 1, 4	37.5	–	50.0	–	68.8

latency to the inter-die critical paths on the fixed dies, tighter timing constraints are added to the dies that may be retimed. This makes retiming more difficult on the dies, but ensures that the added delays on the fixed dies are taken into account during retiming. The first column displays which dies are fixed, and the results given are for percentage delay recovery. Table 6.5 shows results from the experiments of Table 6.4 with the addition of the algorithm to move logic between dies in an attempt to improve results. Results are given only for experiments where two adjacent dies are unfixed, as logic cannot be moved to fixed dies.

As can be seen from the results, retiming recovered 100% of the additional GSF bypass mode delay in most cases. The one exception to this is Die 2 of the four-die stack under die-level retiming.

For the benchmark circuit used in this experiment, retiming becomes more difficult as the circuit is partitioned across more dies. During both die-level and stack-level retiming, slack can only be redistributed from within a die, because logic and registers will not be moved between dies. As the circuit is partitioned between more dies, there are fewer paths on each individual die from which to take excess slack. Furthermore, more paths will cross die boundaries and be subject to additional delay after GSFs are added during DfT-insertion. This increased difficulty of retiming is reflected in the lower recovered delay of Die 2 of the four-die stack.

Table 6.1c and the results of Tables 6.4–6.6 also demonstrate the greater effectiveness of stack-level retiming versus die-level retiming with regard to slack redistribution. For example, if the DES four-die stack were to be assembled after die-level retiming, an additional delay would exist on some of the paths that cross into Die 2 from Dies 1 and 3. This delay would be no worse than 40% of the delay of a GSF in bypass mode. This delay exists because not enough positive slack paths are present on Die 2 or because the extra slack present on Die 2 could not be distributed to TSV paths that violated timing constraints during retiming. During

Table 6.5 A comparison of delay recovery with fixed dies and the additional use of the logic movement algorithm for the four-die FFT logic circuit with two fixed dies

Delay recovery when retimed for:					
Fixed dies	Die 0	Die 1	Die 2	Die 3	Complete stack
Dies 0, 1	–	–	75.0	82.5	100
Dies 2, 3	50.0	75	–	–	71.9

Table 6.6 A comparison of delay, area, and pattern count results for die- and stack-level retiming for the OR1200 processor partitioned across two dies

	Die 0	Die 1	Complete stack
% Delay recovered	37.5	50	92
% Area overhead post-wrapper insertion	6.1	4.6	7.2
% Area overhead post-retiming	5.2	4.4	6.3
% change in pattern count	0.1	0.6	2.9

stack retiming, Die 2 would present the same difficulties to the retiming algorithm. However, because all dies are retimed simultaneously and complete paths that cross dies are known, slack could be redistributed on the other dies in the stack to make up for the restrictions of Die 2.

For the DES benchmark and die partitioning used in this chapter, the area overhead of DfT-insertion ranged from 16.3% for Die 0 of the two-die stack to 35.6% for Die 2 of the four die stack. Area overhead decreased with retiming, as the retiming algorithm retimes for area minimization after clock period minimization. The area overheads are relatively large because with die partitioning many paths crossed dies, and so many TSVs were introduced to the circuit. Compared to the number of cells on each partitioned die before DfT-insertion, the number of GSFs added after DfT-insertion accounted for a large number of the total number of cells. As can be seen from the FFT benchmark, which is much larger and therefore has a larger cell to TSV ratio, the area overheads were significantly lower, with 2.5% being the highest area overhead for Die 2 of the four-die FFT stack.

As the number of dies in a partition increased, the number of cells on each die before DfT-insertion decreased with an often corresponding increase in TSV count. Thus, the area overhead of DfT-insertion generally increased as the stack became larger. The area overhead was also generally shown to be worse for the inner dies of a stack, as these require GSFs both on TSV paths and on paths that contain face-side pads that will be bonded to TSVs on another die during stack assembly.

To illustrate this effect, consider the DES circuit evenly distributed among dies in a four die stack. In this case, each die would have about 6,500 cells. Assume that each die has 500 connections with each adjacent die. The Die 0 and Die 3 of the stack would each need 500 GSFs, or 8% of the cells on the die. In comparison, Die 1 and Die 2 would each need 1,000 GSFs, which is over 15% of the cells on the die. The area impact of GSF insertion would be significantly less if the dies themselves were more complex, with a higher cell count relative to the number of TSVs per die. For example, if a die contained one-million cells and 10,000 TSVs, then adding GSFs to a die would only account for 1–2% of the number of cells on the die, depending on whether or not the die were in the middle of the stack.

Tables 6.4a shows that, generally, there is enough slack present in the FFT circuit such that three of the four dies are capable of recovering their own timing overhead as well as that of the fixed die. This is not the case when two dies are fixed, as seen in Tables 6.4b, although better stack-level retiming results hold true in this case. Furthermore, the logic-movement algorithm to push logic cells between dies can be used to improve die-level retiming under two fixed dies as seen in Table 6.5. In the case of Die 1 when Dies 2 and 3 are fixed, there is a delay recovery improvement from 62.5% recovery without logic movement to 75% recovery with logic movement, or a 16.7% increase in delay recovery. In the case where Dies 2 and 3 are fixed, logic movement has no effect. Utilizing the logic-movement algorithm to push logic cells between dies can improve die-level retiming under two fixed dies as seen in Table 6.5. In the case of Die 1 when Dies 2 and 3 are fixed, there is a delay recovery improvement from 62.5% recovery without logic movement to 75% recovery with logic movement, or a 16.7% increase in delay recovery. In the case where Dies 2 and 3 were fixed, logic movement had no effect.

This example of logic redistribution provides insight into when logic redistribution can improve results and how much improvement can be expected. As demonstrated by Table 6.5, stack-level retiming, in general, produces better results than die-level retiming even with logic redistribution, and logic redistribution does not change stack-level retiming results. This is because stack-level retiming already considers the complete inter-die paths, so logic redistribution will never result in delay recovery improvements (or delay recovery reductions), and at best die-level retiming can only match stack-level retiming.

Additional insights are gained when considering the effect of redistribution with Die 0 and Die 1 fixed, or Die 2 and Die 3 fixed. In the former case, no improvement is seen as the critical paths contain no logic that could potentially be moved. In the latter case, there is improvement in the delay recovery on Die 1, as some logic is moved to Die 0, which contain some additional slack to be provided to the path. However, there is no improvement on Die 0 because its most critical path has no movable logic. Therefore, it is demonstrated that logic redistribution is useful only when two conditions are satisfied—there is movable logic on a die's most critical path, and there is additional slack on the adjacent die's portion of that path.

Delay recovery is also more difficult in a circuit with modules that cannot be retimed, as shown in Table 6.6 with the OR1200 results. Although die-level retiming provided less delay recovery compared to other benchmarks, with 37.5% recovery for Die 0 and 50% recovery for Die 1, stack-level retiming still provided good results, with an overall 92% delay recovery.

The effect of DfT-insertion and retiming on stuck-at fault pattern count was found to be negligible. In some cases, such as Die 0 of the two die DES stack or Die 2 of the three die DES stack, this resulted in fewer test patterns after retiming. In other cases, such as Die 1 of the two die DES stack or Die 0 and Die 1 of the three die DES stack, retiming incurred a small increase in pattern count. It should be noted that when scan insertion is performed for the complete stack, scan chains are allowed to span multiple die layers. This is not the case when scan insertion is performed on a per-die basis. Thus, there is a significant difference in the number of test patterns produced after stack-level scan insertion when compared to die-level scan insertion. Variability is seen in pattern count because registers can be added/removed as well as moved throughout the circuit during retiming, there can be significant changes in controllability/observability during scan test. This can have a wide range of impact on ATPG, and this impact produces the pattern count changes.

Similar conclusions can also be drawn for path delay test patterns, as evident in Table 6.3. The effect of retiming on SDQL was generally greater in the relatively small DES stacks than the much larger FFT stacks. For the DES stack, the worse-case change in SDQL is a 6% reduction, while for the FFT stacks at worst a 1% reduction is seen. The greater volatility of SDQL for the smaller stack implies that, when dies are larger, most die-internal paths are untouched by retiming. Since there are far fewer die-internal paths in the DES benchmark, changes in paths at or near the die boundary make up a change in a larger percentage of the total die paths. Changes in SDQL in the same design depending on stack size is due to differences in logic partitioning, which alters which paths are targeted for test by the statistical ATPG model.

Run times for the retiming algorithm were generally in minutes per die, but were longer for a complete stack and larger benchmark. For example, the run time for the complete four-die DES stack was 12.4 min. For each die, the run times (in minutes) were 3.5, 2.7, 2.6, and 2.2 for Die 0, Die 1, Die 2, and Die 3, respectively. For the FFT four-die benchmark, run times in minutes for die-level retiming were 9.4, 10.6, 10.2, and 9.1 for Die 0, Die 1, Die 2, and Die 3, respectively, while stack-level retiming required 41.5 min.

6.2.4 Summary

Section Summary

- Retiming performed after die-wrapper boundary register insertion can be used to recover the delay overhead added by DfT insertion to inter-die paths.
- Retiming can be performed at die- or stack-level depending on design constraints, with stack-level retiming providing the best retiming results.
- Generally, retiming can recover 100% of the delay overhead in non-modular logic design, and above 50% of the delay overhead in stacks with fixed dies or dies with fixed modules.
- Logic redistribution can improve retiming results in certain designs, although the method is limited due to constraints on adding or removing TSVs.

6.3 Conclusions

The methods and results discussed in this chapter show that retiming can be used to recover the delay added to circuit paths that cross die boundaries during die DfT-insertion. Retiming has been demonstrated at both the die-level and stack-level, with stack-level retiming providing an upper limit on die-level retiming. Retiming results on a DES circuit and an FFT circuit partitioned into two, three, and four die 3D benchmarks are provided. In most cases retiming can recover 100% of the delay added by die DfT-insertion when all logic and dies are unfixed and in cases where a quarter of dies are not unfixed. It is further shown that, for modular benchmarks and benchmarks in which half of the dies are not unfixed, stack-level retiming outperforms die-level retiming in terms of delay recovery. However, a logic redistribution algorithm can be utilized to improve die-level retiming results in some cases. It has also been demonstrated that test pattern counts are not significantly impacted by DfT-insertion or retiming.

Chapter 7
Post-Bond Test Wrappers and Emerging Test Standards

7.1 Introduction

Throughout this book, methods and architectures have been discussed for performing pre-bond and post-bond test in 3D SICs. Many of these methods, such as those discussed in Chaps. 4 and 5, were designed to be compatible with emerging test standards. This chapter will explore the emerging standards with regard to stacked dies and test interfaces and their associated test implications. Section 7.2 examines a die test wrapper, based on the IEEE 1500 and IEEE 1149.1 [20, 102] test standards, that is currently being developed by the IEEE P1838 workgroup [103]. Section 7.3 provides an overview of the JEDEC JESD-229 [104] standard developed for memory-on-logic stacks and how the test wrapper described in Sect. 7.2 can be extended for testing a stack that utilizes the JEDEC framework.

It is assumed that the reader has some knowledge regarding test standards for 2D circuits, e.g. the IEEE 1500 standard and the JTAG 1149.1 standard. The 1500 standard, whose most basic implementation is shown in Fig. 7.1, is utilized to present a standardized test interface for individual test modules on a die. Although Fig. 7.1 shows only a single test module, multiple modules can be chained together with a test access mechanism (TAM) to provide test I/O for all wrapped modules via external pins. The 1500 standard consists of wrapper boundary registers (WBRs) which capture signals from the module I/O and are connected together as a scan chain for test I/O. These provide for both die internal and die external tests.

A wrapper instruction register (WIR) is programmed from the wrapper serial ports, such as the wrapper serial input and output (WSI and WSO, respectively) ports. Other serial inputs include the wrapper clock and reset ports, the instruction register selection port, and shift and capture ports. The WIR places the wrapper into various test states. A bypass port is available to skip a test module on the TAM to send and receive test data from a module elsewhere. Optional components can be added to the wrapper, including serial ports for the rapid routing of test data.

B. Noia and K. Chakrabarty, *Design-for-Test and Test Optimization Techniques for TSV-based 3D Stacked ICs*, DOI 10.1007/978-3-319-02378-6_7,

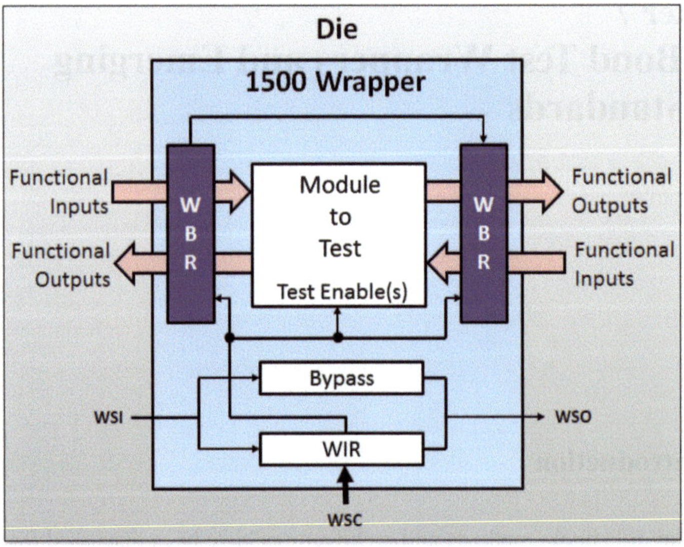

Fig. 7.1 The required components of the IEEE 1500 standard

The 1149.1 JTAG standard defines a standard test interface for multiple test modules on a printed circuit board. It defines four mandatory and one optional test pin. Mandatory I/O include a test data in (TDI) signal, test data out (TDO), test mode select (TMS), and the test clock (TCK). An optional test reset (TRST) may also be included. The JTAG standard utilizes boundary scan testing at the board level, similar to the function of the WBR in the 1500 standard. These boundary registers allow for both internal test (INTEST) and external test (EXTEST).

The die wrapper that will be introduced in Sect. 7.2 presents a similar test interface building on both the JTAG and IEEE 1500 for dies in a stack. The die wrapper is compatible with both standards and may be designed with either architecture in mind. A standard for die testing provides the modularity required to enable the integration of dies from various sources into a functional and testable stack. Although the 3D integrator may have no knowledge of the test architectures used on a die, a vendor can provide test patterns and a standardized interface such that the integrator would be capable of testing the die.

The JEDEC standard for Wide-I/O Mobile DRAMS (JESD-229) was developed to define the interface between a memory and logic die in a 3D stack. It defines a 512 bit interface that can provide significant bandwidth at relatively low power. As the first standard interface of its kind for 3D stacks, backed by the JEDEC Solid State Technology Association, it is expected to play a significant role in the design of future memory-on-logic stacks. Section 7.3 will briefly introduce the JEDEC standard and then extend the die wrapper design with additional functionality to create a standardized test interface for interconnect testing of memory-on-logic stacks.

7.2 Die Wrappers for Standard Test Access of 3D Stacked ICs

In order for a die-level wrapper to be utilized, every die in the stack should be wrapped in accordance with the standard. The die wrapper supports a reduced-bandwidth pre-bond test mode, post-bond testing for both partial and complete stacks, and board-level interconnect testing. The die wrapper is compatible with both the 1500 and JTAG 1149 standards, and is therefore modular in design. In other words, each die, its embedded test modules, TSV inter-die interconnects, and external pins, can all be tested separately. In this way, the scheduling of pre-bond, post-bond, and board-level tests can be flexible depending on the tests needed for a particular manufacturing flow. Although the die wrapper standard is still under development through the IEEE P1838 test work-group, the die wrapper discussed in this section will be referred to as the P1838-style standard wrapper for convenience.

The P1838 wrapper assumes that external connections to the stack are available only on either the top or the bottom of the stack. While it is possible to have I/O connections elsewhere in the stack, for example via wire-bonding, it is likely in the immediate future that I/O will be available only through the bottom of the stack. In this section, it is assumed that this is the case in order to simplify explanations, although by switching references from the bottom of the stack to the top of the stack the reader can understand how the wrapper would be implemented were I/O available through the top of the stack.

The P1838 wrapper is designed to accommodate a variety of pre-bond and post-bond test scenarios. For example, after the fabrication of a die one company may want to perform pre-bond KGD test, including the test of all internal modules, intra-die circuitry, and TSV test (although pre-bond TSV testing is not explicitly supported by the wrapper, Chaps. 3 and 4 provide possible solutions). The company may then ship the good dies to a second company for integration into a stack, and the second company would like to perform post-bond testing of the partial and complete stack, perhaps retesting the internal modules of each die as well as the TSV interconnects between dies. The die wrapper is further integrated with the 1149.1 standard for board-level test.

Each die in the stack is assumed to be equipped for scan testing, i.e. scanable digital logic, BIST circuits, etc. The wrapper interfaces with the internal scan chains, test control architectures (such as 2D TAMs), compression circuits, etc. In order to accommodate the wrapper and its functions, the addition of TSVs to the design may be necessary to allow for communication between the dies during test. These dedicated test TSVs are referred to as *test elevators* and will be discussed in more detail later.

Due to the availability of external pins only on the bottom (or top) of the die, all test signals must be routed to and from each die in the stack via dies lower in the stack. In other words, all test control signals and test data must be routed through the bottom die of the stack. When these signals are routed to or from the die for which they are intended without moving further in the stack, it is called a *test turn*.

The die wrapper is tier-neutral in that a die equipped with the wrapper may be placed anywhere in a stack. Furthermore, the wrapper does not limit the number of dies that can be in a stack.

Similar to test standards for 2D circuits, the P1838 wrapper is scalable as required by the designer. A one-bit serial TAM is required, with an optional multi-bit parallel TAM. The serial TAM is utilized for debug and diagnosis. Similar to the 1500 serial ports, it provides a low-cost, low-bandwidth method of loading instructions for test configuration and routing test data. The serial ports can be used after the stack is integrated onto a circuit board. The optional parallel TAM provides a method for high-volume production testing. While more costly to implement, it can significantly reduce production test time.

Due to the modularity of the P1838 wrapper, various tests can be performed at different times or not at all—there is no requirement to test the stack as a single entity. All of the possible interconnect tests between dies and the dies themselves are considered as separate test insertions. Each die can be made up of any number of embedded test modules, and these too may be treated as separate entities in regard to test. The benefit of such a modular approach to test is in ease of integration of IP modules or dies, the ability to optimize test for different fault models depending on the circuit under test, and the freedom in designing and optimizing the best test flow for any given stack.

7.2.1 The Die Wrapper Architecture

Each die in a stack is equipped with a die wrapper, and the wrappers of each die work together to enable test as shown in Fig. 7.2. Figure 7.2 provides a conceptual overview of the die wrapper in a three-die stack soldered onto a circuit board, where each die in the stack is wrapped. The pins on the bottom die of the stack provide both functional I/O and, in this example, test inputs. Two types of TSVs exist between each die—FTSVs are functional TSVs, while TTSVs are dedicated test TSVs for use by the die wrappers, which are also referred to as *test elevators* or *TestElevators*. Each die contains some number of test modules that are individually wrapped via the 1500 wrapper, and each die contains its own 2D TAM connecting the internal test modules to one another and the modules to the die wrapper. The modules do not need to be wrapped—they can also include test compression, BIST, or other testable modules. An 1149.1 compliant test-access port (TAP) exists on the bottom die to enable board test.

The die wrapper added to each die in the stack is the additional DfT architecture that makes up the P1838 standard. The wrapper consists of serial I/O for configuring the wrapper test mode and possibly parallel test interfaces as well. Arrows show the movement of test data throughout the stack, and test turns exist on each die to accept data from and return data to the I/O pins on the bottom die. Large, dedicated probe pads are shown to enable a reduced-bandwidth pre-bond test mode. These large pads provide a landing point for probe needles to individually contact TSVs for

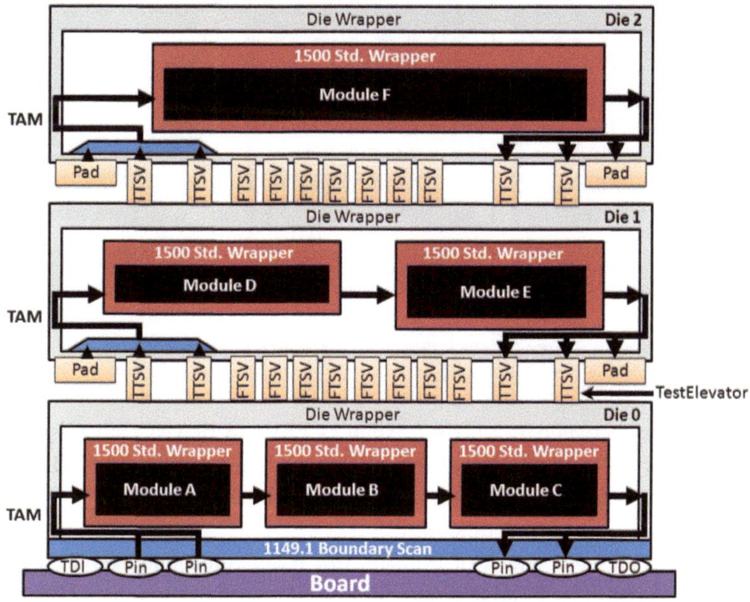

Fig. 7.2 A conceptual example of a three-die stack using die wrappers

test application during pre-bond test. Architecture extensions discussed in Chaps. 4 and 5 provide compatible alternatives to these probe pads. The test elevators are used to route test signals up and down the stack from the bottom I/O pins. A 3D TAM, which consists of the test turns, test elevators, and a control mechanism for setting the individual die wrapper test modes and optionally the embedded module test modes also exists as part of the die wrappers.

7.2.2 1500-Based Die Wrapper

The die wrapper can be designed to interface with either the 1500 or 1149.1 standard. Both will be discussed, starting with the 1500 implementation shown in Fig. 7.3 for a three-die stack. Similar to 1500, it can have two test access ports—a mandatory single-bit serial port with a wrapper serial in (WSI) and wrapper serial out (WSO) port used for providing instructions to the wrapper and for low-bandwidth test, and an optional parallel access port for high-bandwidth test. The parallel port can be of arbitrary size, depending on the needs of the designer. The bits of the wrapper instruction register (WIR), combined with the wrapper serial control (WSC) signals, determine the mode of operation that the wrapper is in at any given moment. The wrapper boundary register (WBR) is used to apply test patterns and capture responses for both internal tests to the die itself (*intest*) and external

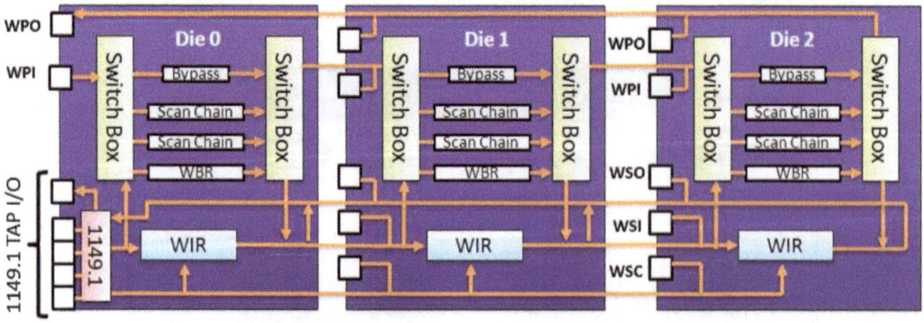

Fig. 7.3 An example of a three-die stack utilizing 1500-based die wrappers

tests for circuitry between dies (*extest*), such as TSVs. A bypass path exists to route test data past a die without testing the die or needing to utilize the WBR. Intest, extest, and bypass comprise three of the possible modes of operation in P1838, and are analogous to their 1500 counterparts.

As can be seen in the stack of Fig. 7.3, the WSC control signals are broadcast to all the die WIRs. The serial and parallel test buses are daisy-chained through the stack. The highest die in the stack does not utilize test elevators, as there is no die above it. All external I/O pins are on the bottom die, as is an 1149.1 TAP controller to provide for board testing. The serial interface of the die wrapper on the bottom die is connected to the TAP. The only additional pins required for the P1838 wrapper are those for the standard JTAG interface as well as optional parallel ports.

There are four significant features of P1838 that differ from 1500 and are unique to 3D SICs. These features are:

- *Test turns*—Modifications are made to the standard 1500 interface, which exists on the bottom side of each die and is made up of WSC, WSI, WSO, WPI, and WPO. Pipeline registers are added at the output ports (WSO, WPO) to provide an appropriate timing interface between dies or the stack and the board.
- *Probe pads*—All dies on the stack except for the bottom die, which is already equipped with the I/O pin locations for the stack, have oversized probe pads added to some of their back-side TSVs or face-side TSV contacts. These provide a location for current probe technologies to touchdown on the die and individually contact each probe pad. In P1838, these probe pads are required on WSC, WSI, and WSO, which represent the minimal interface for the die wrapper. Probe pads may also be added on any or all of the WPI and WPO pins as necessary. If fewer probe pads are added for pre-bond test than there are parallel access ports for post-bond test, a switch box is utilized to switch between a low-bandwidth pre-bond test mode and a higher-bandwidth post-bond test mode. The state of the switchbox is controlled by the WIR. While the probe pads are currently necessary for testing through the die wrapper in the P1838 standard, previous chapters have discussed methods for providing a test interface that requires few probe pads while providing for pre-bond TSV and scan test.

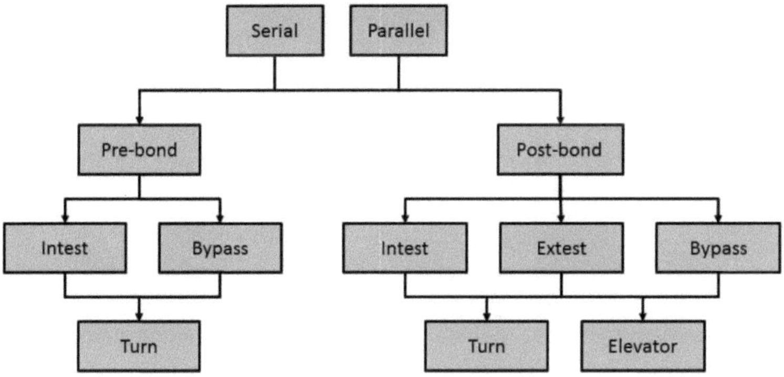

Fig. 7.4 A diagram of the possible modes of operation for the P1838 die wrapper

- *Test elevators*—Additional, dedicated test TSVs for the die wrapper are necessary for routing test data and instructions between dies. These TSVs are referred to as test elevators.
- *Hierarchical WIR*—In the 1500 standard, each embedded test module on each die is equipped with its own 1500-compliant wrapper. In order to provide instructions to these internal wrappers, a hierarchical WIR is necessary. In order to load all WIRs, the WIRs are chained together similarly to scan chains. The length of the WIR chain depends on the number of dies in the stack, the number of 1500-compliant test modules per die, and the summed length of the WIR instructions. In P1838, the die wrapper WIRs are equipped with an extra control bit to bypass the WIRs of the embedded test modules on the die in order to only load die wrapper WIRs.

Figure 7.4 shows the possible operating modes of the P1838 wrapper. To read the diagram, start from either the serial or parallel test mode and follow a path to the end of the diagram. Each path that can be made is a possible operating mode. For example, several possible modes of operation include *Serial-PrebondIntestTurn*, *ParallelPrebondIntestTurn*, *ParallelPostbondExtestTurn*, and *SerialPostbondExtestTurn*. There is a total of 16 operating modes, with 4 pre-bond modes and 12 post-bond modes.

Each die may be in a different operating mode, depending on what is tested within the stack at any given time. For example, any or all dies may be tested simultaneously. Likewise, any or all interconnect tests between dies can be performed simultaneously. To give an example, consider a four-die stack in which the interconnects between Dies 2 and 3 are tested in parallel with the internal circuitry of Die 4. All of these tests take place utilizing the parallel access ports. In this example, each die except for Dies 2 and 3 will be in a different operating mode. Die 1 will be in *ParallelPostbondBypassElevator* mode, as it is being utilized as a bypass to route test data up and down the stack. Die 2 and Die 3 are placed in *ParallelPostbondExtestElevator* mode, as they are performing their external test on

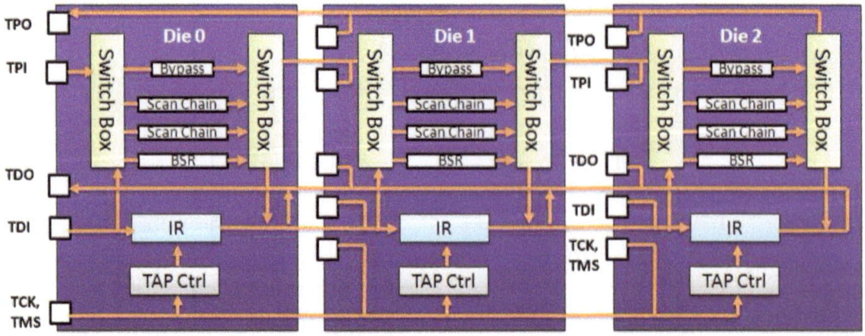

Fig. 7.5 An example of a three-die stack utilizing 1149.1-based die wrappers

the TSVs between them as well as routing test data further up the stack. Die 4 is in *ParallelPostbondIntestTurn* mode, performing its internal module tests and turning test data back down the stack.

7.2.3 JTAG 1149.1-Based Die Wrapper

The 1149.1 standard is utilized to provide a test wrapper for chips on a circuit board. It can be extended across all of the dies in the stack in P1838 to provide test functionality through die wrappers. There is significant overlap in design between the 1500- and 1149.1-style die wrappers, including test turns, probe pads, test elevators, and operating modes. The differences between the two die wrappers are the focus of this section.

Die wrappers in the style of the 1149.1 standard are shown for a three-die stack in Fig. 7.5. The 1149.1 standard only has a serial test access mechanism through the test data in (TDI) and test data out (TDO) ports. In order to provide high-bandwidth test for the stack, additional parallel ports (TPI and TPO) can be added to the die wrapper. In place of the WSC port utilized in the 1500-style die wrapper, a two-bit JTAG control port is utilized. This port consists of the test clock (TCK) and test mode select (TMS) ports, and an optional third port for test reset (TRSTN). The reset port is not shown in the figure. In order to provide the necessary control signals to the instruction register (IR) of the die wrapper, a 16-state finite state machine (FSM) called the TAP controller (TAP Ctrl) is utilized. This can be stepped through via the TMS signal to provide the appropriate instructions for the wrapper.

One benefit of using the 1149.1-style wrapper is that the hierarchical WIR is easily implemented without additional design effort. This is because the 1500-compliant wrappers on embedded test modules are already implemented in a hierarchical relationship to the 1149.1 instruction register that is included on each die. There further exists a wide range of optional uses of the JTAG interface,

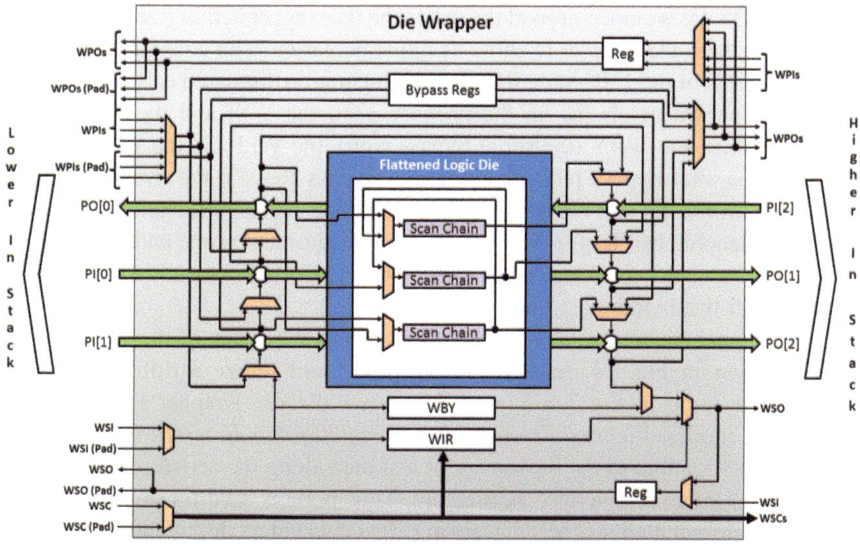

Fig. 7.6 An example implementation of the 1500-style die wrapper for a flat die

including software debug and emulation. Having a die wrapper based on the 1149.1 standard allows for these additional features to be easily integrated into the stack if desired.

7.2.4 P1838 Die Wrapper Example Implementation

This section provides an example implementation of a 1500-style die wrapper [105]. Though only a 1500-style implementation is provided, an 1149.1-style implementation can be similarly developed and has significant overlap with the example in this section. To reduce the complexity of the implementation, it will be shown for a "flat" die, or a die that does not include embedded test modules. The die is treated as a single, scanable entity for test. The number of probe pads for pre-bond test will be equal to the number of test elevators for the die. A wrapper for a modular die, with the exception of a hierarchical WIR structure, is similar to the example in this section. Therefore, the implementation of a hierarchical WIR will also be discussed for completeness.

Figure 7.6 shows the implementation of the 1500-style die wrapper for a flat die. The die contains three functional primary inputs (PIs) and three primary outputs (POs). In the stack, this die will be bonded to two other dies. The primary I/O on the left side will be connected to a die lower in the stack, and the primary I/O on the right side will be connected to a die higher in the stack. The flat die has three internal scan chains.

The P1838 die wrapper, around the entire die, is exaggerated in size compared to the internal die logic in order to show its implementation. The wrapper contains all of the elements of the 1500-style P1838 implementation discussed earlier, including the WBR (shown as circles on the primary I/O), the WSI and the WSO serial ports, the WIR, the WBY (the serial bypass path), the WPI and the WPO parallel ports, the parallel bypass path (denoted as "Bypass Regs"), the oversized probe pad connections, the test elevators, and the pipeline registers included for timing purposes (denoted by "Reg"). The parallel test elevator and probe pads number the same, so no switch box is necessary to switch between a low-bandwidth pre-bond test and high-bandwidth post-bond test.

The wrapper implementation is capable of being set to any of the 16 operating modes shown in Fig. 7.4. Each operating mode will utilize a different path of the wrapper logic in Fig. 7.6. Figure 7.7 shows the die wrapper in two of the operational modes. Figure 7.7a shows the *ParallelPrebondIntestTurn* mode, with arrows corresponding to the movement of test data along the activated paths of the wrapper and internal logic test architecture. This mode is used for high-bandwidth test of the internal die logic before stacking. Test data enters through the WPI probe pads, where it is latched into the WBRs and applied to the internal scan chains. Test responses are then turned out of the internal logic through the WBY, the pipeline registers, and then to the test equipment via the WPO probe pads.

Figure 7.7b shows the *SerialPostbondExtestElevator* die wrapper mode. This is utilized for low-bandwidth testing of the interconnects between this die and other dies in the stack after bonding. Test data enters through the single-bit WSI port, where it is latched in the WBR. The one-bit WSO on the right of the die is used to shift extest data to the die higher in the stack, which must also be in extest mode to perform interconnect test. From the WBR, the data is applied to the interconnects and test responses are latched. The test responses are then shifted out of the path through the higher die in the stack by use of the WSI port on the right side of the die and the WSO port on the left in order to route the data out from the lowest die in the stack.

In order to configure the die wrapper into the desired operating mode, multiplexers are utilized; these are controlled by the WSC signals and the WIR. These multiplexers perform a variety of functions, including switching between parallel or serial modes, determining whether the serial ports are used for loading the WIR, the WBR, or the WBY, and so forth. Many of the control signals are not shown in order to preserve the readability of the picture.

A hierarchical die, in which there are multiple embedded test modules, would have a similarly designed die wrapper with several modifications. The most significant modification is the need for a hierarchical WIR between the die wrapper and the embedded test modules. An example implementation of a hierarchical WIR is shown in Fig. 7.8. The figure shows only the first internal test module on the internal WIR chain, because the implementation is the same regardless of the number of embedded WIRs. The die-level WSC signals are sent to the first module WIR except for WRSTN, which is AND multiplexed with an enable signal from the die wrapper WIR. The die wrapper WIR asserts the enable signal when the module

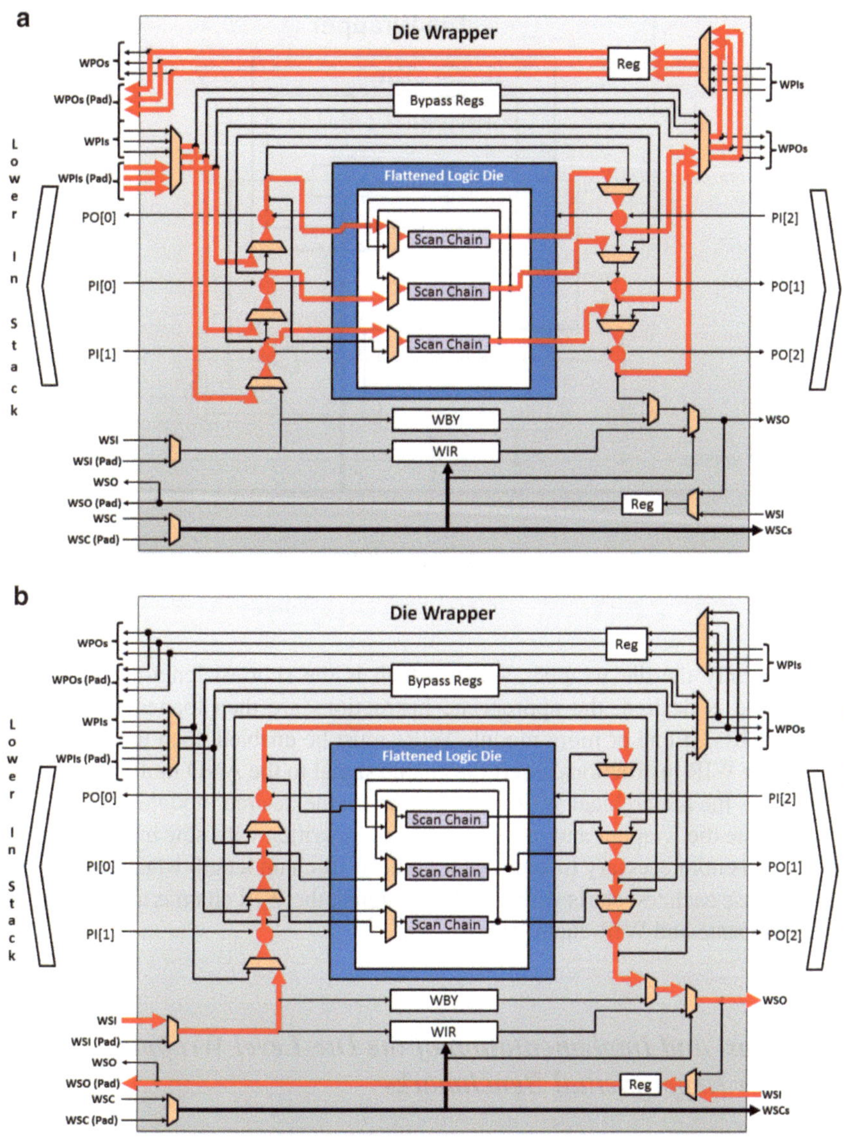

Fig. 7.7 An example of the logic and paths utilized by the die wrapper for (**a**) *ParallelPrebondIntestTurn* and (**b**) *SerialPostbondExtestElevator* test modes

WIRs should be enabled. When the enable signal is asserted, the multiplexer shown in Fig. 7.8 adds the module WIR to the WIR chain.

There are three operations that can be performed when using a hierarchical WIR. These are a WRSTN reset followed by zero to two instruction loads as needed. The WRSTN sets all WIRs into their functional mode. This also resets the WIR chain

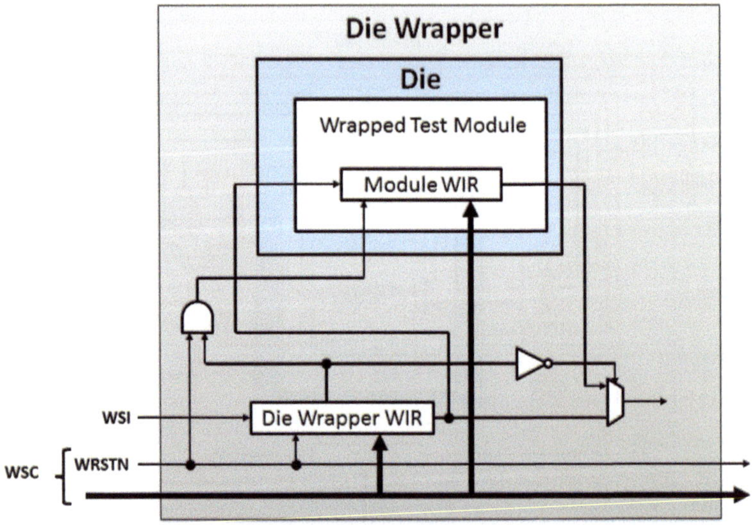

Fig. 7.8 An example implementation of a hierarchical WIR for a 1500-style die wrapper for a die with multiple embedded test modules

to include only the die wrapper WIRs, which is the shortest length WIR chain. If test mode is desired, the appropriate instructions are then loaded into all die wrapper WIRs. If one or more module WIRs must be enabled, then the associated die wrapper WIR instructions assert the enable signal to the AND multiplexer. This reconfigures the hierarchical WIR chain to include the desired module WIRs. After this point, the die wrapper and module WIRs are rewritten with new instructions. In this way, it is not necessary to keep track of the WIR chain length when re-ordering tests. Because each test starts with a WRSTN signal, the WIR chain is always placed in a known state and WIR chain length.

7.2.5 Cost and Implementation of the Die-Level Wrapper for Experimental Benchmarks

It is important to quantify the area overhead of the P1838 wrapper architecture in order to determine if the architecture is feasible from a cost perspective. The P1838 wrapper requires the addition of TSVs in the form of test elevators, oversized probe pads for test access, and additional logic circuitry. The 1500-style die wrapper requires 8 probe pads for the necessary WSC and serial ports as well as some optional number of probe pads depending on how many bits, if any, of the parallel port are desired to be active in pre-bond test. The 1149.1-style die wrapper requires four probe pads for TDI, TDO, TCK, and TMS signals, plus some number of additional pads for the parallel ports as desired. Note that additional probe pads are needed for other necessary signals such as power, ground, clocks, and so forth. In addition to the probe pads, it is necessary to add some extra number of TSVs to the design. For the 1500-style wrapper, eight TSVs plus TSVs for all parallel ports are required. Similarly, for the 1149.1-style wrapper, four TSVs plus TSVs for the number of parallel ports in the design are needed.

The area cost of the additional logic circuitry needed for the die wrapper is dependent on three components. First, there is a fixed cost associated with the WIR, WBY, and a number of multiplexers. Second, there is a variable cost associated with the number of functional I/Os on the die. This cost scales linearly with the number of I/O and includes the cost of the WBR registers. Lastly, there is a variable cost associated with the number of scan chains in embedded modules. This cost scales linearly with the number of scan chains and includes the cost of the multiplexers needed for concatenating scan chains.

The area cost A for the P1838 die wrapper can be estimated as:

$$A = F_{cost} + (\#IO \cdot IO_{cost}) + (\#SC \cdot SC_{cost}), \tag{7.1}$$

where F_{cost} is the fixed cost of the WBR and associated logic, SC_{cost} is the cost per scan chain, and IO_{cost} is the cost per functional I/O. The variables $\#IO$ and $\#SC$ represent the number of functional I/O and embedded scan chains, respectively.

The authors of [105] created a tool flow for adding a P1838 die wrapper to a circuit. The flow begins with a gate-level netlist of the die to which a commercial EDA tool adds a conventional test wrapper. The conventional wrapper is then modified by hand to create the die wrapper. ATPG is used to verify the design, and area overhead is determined by reporting the gate area of the wrapped design versus the unwrapped design.

The authors utilized their flow for three benchmark circuits from the ISCAS'89 benchmarks [106]—s400, s1423, and s5378. Each of these circuits were utilized separately as dies to be wrapped. They were mapped to the Faraday/UMC 90 nm CMOS standard cell library. Table 7.1 shows the area overhead for each of these circuits. Column 1 shows the name of the circuit, Column 2 shows the estimated area overhead for the wrapper as determined using Eq. 7.1, Column 3 shows the actual area overhead as determined by the wrapper insertion flow, and Column 4 shows the area overhead of the additional DfT features as a percentage, calculated by dividing the wrapped area by the unwrapped area.

Costs were determined by layout using the Faraday/UMC 90 nm standard cell library. Costs were calculated as $F_{cost} = 432\ \mu m^2$, $IO_{cost} = 36\ \mu m^2$, and $SC_{cost} = 63\ \mu m^2$. As can be seen from the Table, the estimated and actual areas after wrapper insertion are very close, so the estimate is considered accurate. The designs from the ISCAS'89 benchmarks are very small, with s400 consisting of 186 cells, s1423 having 734 cells, and s5378 having 2,961 cells. Due to their small size, the area overhead of wrapper insertion is significant. However, the area overhead drops rapidly with increasing circuit size as the die wrapper takes up less area compared to the overall design. An industrial circuit would be orders of magnitude larger than the ISCAS'89 benchmarks. To demonstrate the area overhead of the wrapper

Table 7.1 Experimental area overhead of the P1838 die wrapper for several benchmark designs

Circuit	Wrapper area (μm^2)		
	Estimated	Actual	Overhead (%)
s400	945	942	90.5
s1423	1,413	1,411	37.7
s5378	3,645	3,645	31.0
PNX8550	19,332	N/A	0.04

architecture in an industrial circuit, PNX8550 [107] was utilized. The estimated area overhead for this larger design is also shown in Table 7.1. As can be seen for a large design, the wrapper area overhead is negligible at 0.04 % additional overhead.

7.2.6 *Summary*

Section Summary

- A DfT architecture is given for 1500-style or 1149.1-style die-level wrappers to enable a standardized test interface for pre-bond and post-bond testing of 3D SICs.
- Included in the architecture are serial and parallel test access ports, test turns to and from die-external I/O, oversized probe pads for pre-bond test with associated switch boxes, test elevators for relaying test data up and down the stack, and a hierarchical WIR infrastructure.
- Area overhead of the die wrapper is shown to be negligible in industrial-sized designs.

7.3 The JEDEC Wide I/O Standard for Memory-on-Logic 3D Stacks

The JEDEC Wide-I/O standard for DRAMs (JESD-229) stacked on logic defines the interface between logic and memory dies in terms of functional and mechanical properties. The conventional DRAM interface is only 32 bits, whereas the JESD-229 interface consists of 512 bits. Unlike logic circuits which are optimized for performance and power consumption, DRAM tends to focus on reducing the area of DRAM and improving refresh requirements. Because of this, heterogeneous integration, where memories and logic are produced on different dies and then stacked, can produce cheaper stacks with better overall performance. The availability of

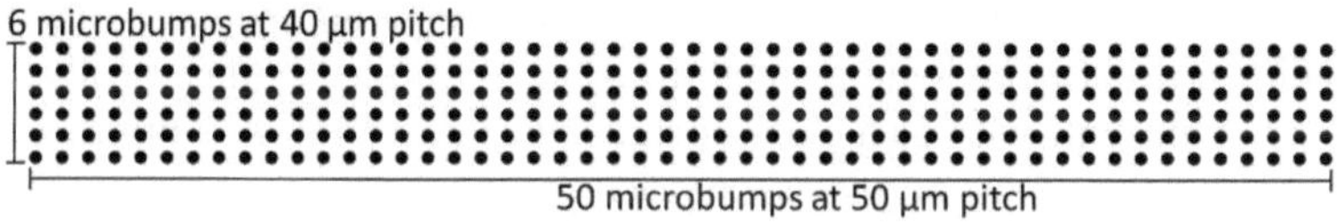

Fig. 7.9 The interface array for a single channel of the JESD-229 standard

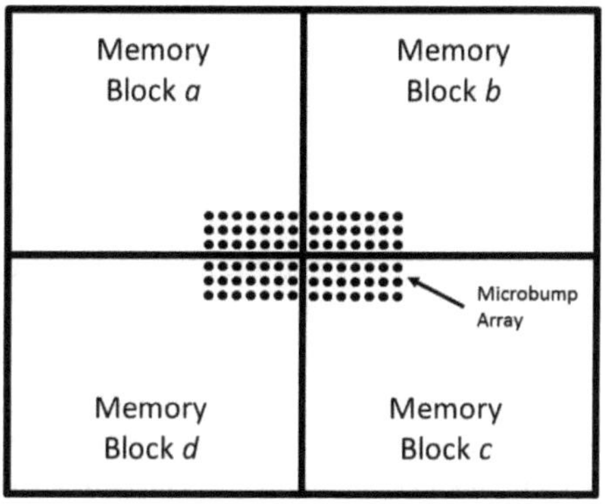

Fig. 7.10 The arrangement of memory blocks and microbump arrays for a single memory die in the JESD-229 standard

TSVs in a 3D stack to provide dense, low-power interconnects benefits memory-on-logic designs. The benefits of using the JESD-229 standard is that, compared to its predecessors such as the LPDDR2 DRAM, it improves power consumption and bandwidth. The JESD-229 standard defines characteristics such as electrical specifications, usage protocols, and ball-out. These contribute to the functional aspects of the technology. Mechanical aspects include locations of pads in an array, array dimensions, and tolerances.

The JESD-229 standard defines four independent memory channels to make up its interface. These are labeled as channel a through d, each with 128 bi-directional bits each, totaling 512 bits across the channels. The JEDEC standard is single data rate with a maximum speed of 266 Mbps. In total, this provides 17 GByte/s of bandwidth between the logic and memory dies. Along with the 128 data bits, each memory channel contains 51 signals for control, address, and clock. There are also shared pads for test control, ground, and power connections.

Mechanically, each pad in the interface is made up of a microbump. There are 300 microbumps per channel for 1,200 contacts across all four channels. The microbumps are placed in regular arrays that are symmetrical for each channel as shown in Fig. 7.9. Each array, one per channel, contains 6 rows of microbumps by 50 columns. The pitch of the microbumps along the rows is 40 μm, and the pitch along the columns is 50 μm. The complete interface for a channel occupies an area of 0.52 mm by 5.25 mm.

The JEDEC standard accommodates up to four memory dies stacked on top of one another. Each die is referred to as a *rank*. Each rank has four blocks of memory as shown in Fig. 7.10. The memory blocks are labeled starting at the top left of the rank and moving clockwise around it, starting at a and ending with d. The microbump arrays for the logic-memory interface are placed symmetrically around the center of the rank, with each block containing its own array. A complete stack of four ranks contains 16 memory dies, but each channel can only access one block at a time. Thus, up to four blocks may be accessed simultaneously. It should be noted that the JEDEC standard does not specify how the dies are stacked. For example, one design may use four dies stacked vertically atop one another with a logic die at the bottom, while another may stack the dies side-by-side on top of an interposer. Furthermore, there is no specification regarding the number of logic dies per stack, only a limit on the number of memory dies.

Probing is necessary to perform pre-bond test of memory dies using the JESD-229 standard. If probing is used, all contacts must be made through the logic-memory interface, as this is the only interface available. Previous chapters have discussed methods and probe technologies that can provide for pre-bond test, and the rest of this section will focus on post-bond testing.

In post-bond testing, functional read and write operations can be utilized via the memory controller on the logic die to test the DRAMs. Memory BIST techniques integrated in the memory controller or elsewhere may be utilized for this purpose. The JESD-229 standard also provides two additional modes that can be used for post-bond memory test. The first mode is the *Direct Access* mode. This allows the testing of all of the DRAMs through a small number of package pins that connect to the DRAM microbumps. The second mode is the *GPIO Test* mode. In this mode, all of the DRAMs can be tested through a small number of GPIO drivers and receivers. These drivers and receivers are part of the circuitry of the logic die that interfaces with the memory.

The next subsection extends the P1838 die wrapper for testing interconnects between the logic and memory dies, as these interconnects cannot be quickly tested by standard memory BIST. The inclusion of a die wrapper will allow interconnect testing to take place separately from other memory tests, and can be useful for partial and complete stack tests. Because testing of all of the DRAM requires a significant amount of time, it may be infeasible to perform it only for interconnect testing. Providing dedicated interconnect testing through the die wrapper requires little time and effort.

7.3.1 Extending the P1838 Die Wrapper for Testing in the JEDEC Environment

While most DRAMs do not usually have boundary scan testing available, DRAMs designed to the JESD-229 standard do. The standard is available for interconnect testing of the JEDEC interface, but does not conform to 1149.1 or 1500 standards. This boundary scan functionality can be largely reused for interconnect test with die wrappers.

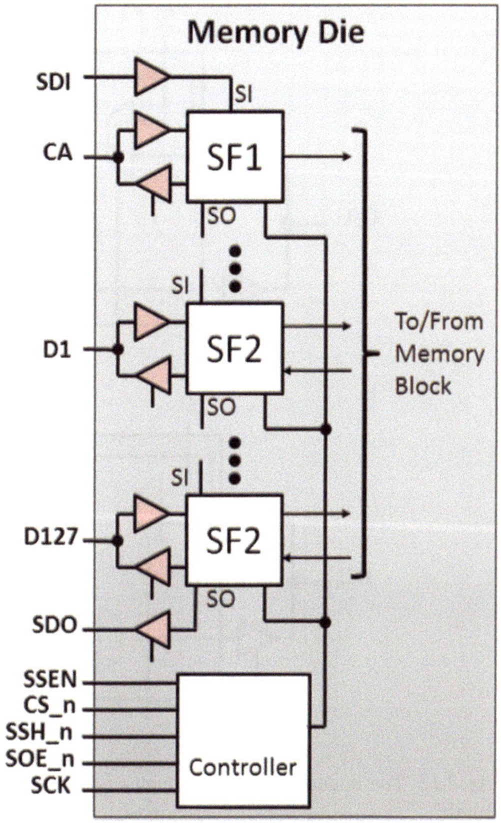

Fig. 7.11 The boundary scan implementation for a memory block in the JESD-229 standard

Figure 7.11 shows the boundary scan implementation outlined in the JESD-229 standard for a memory block. There is a total of 179 functional signals, all of which are latched into one of two kinds of boundary scan flops (SFs). Figure 7.12a, b show the boundary scan flop implementations for SF1 and SF2, respectively. Of the 179 signals, 51 are uni-directional control and address signals that are connected to SF1-type boundary scan flops. The other 128 data signals are bi-directional and are intercepted by SF2-type boundary scan flops. These boundary scan registers provide controllability and observability at the driver and receiver for the TSV interconnect, which allows for external interconnect testing (extest). The input and output signals for scan test—scan-data-in (SDI) and scan-data-out (SDO), respectively—are shared among the ranks of each channel. It is therefore necessary that the drivers for the SDO signals are tri-stateable to be able to select the boundary register of the appropriate rank in a channel at any given time.

A controller is used to operate boundary scan in the JESD-229 standard, and control signals are internal to the DRAM. The signals are generated by the controller itself, which in turn receives signals from a logic die. The signals necessary for the controller are as follows:

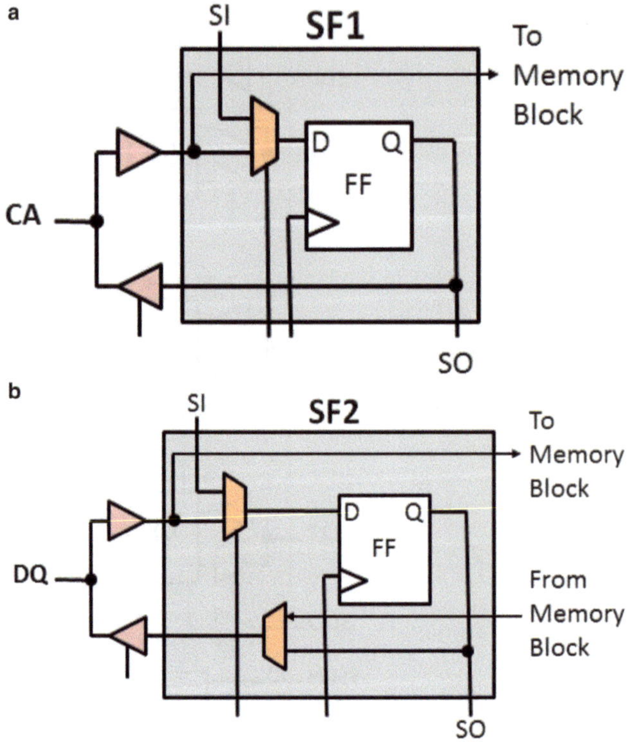

Fig. 7.12 The boundary scan flop implementations in the JESD-229 standard for (**a**) SF1 and (**b**) SF2

1. *SSEN*—SSEN is the one-bit scan enable signal for the DRAM stack. To enable boundary scan, the bit is set high. Otherwise, it is low during functional operation. Should boundary scan of the memory die not be necessary, the SSEN signal is tied to ground.
2. *CS_n*[0:3][a:d]—The CS_n signal, known as the *chip select not* signal, is up to four bits wide per channel in a four-die stack. The signal is used to activate up to one memory block per channel and is active low. It is utilized in both functional and test modes.
3. *SSH_n*[a:d]—SSH_n is the *scan shift not* signal. It is one bit wide per channel and enables shifting of the boundary scan chain. When low, the SSH_N signal activates the shifting of test data through the boundary scan chain. When high, the boundary registers capture inputs from their associated microbumps.
4. *SOE_n*[a:d]—The SOE_n signal, known as the *scan output enable not* signal, is one bit wide per channel. If low, it enables the tri-state driver of the SDO signal for the memory block in the channel selected by the CS_n signal.
5. *SCK*[a:d]—The SCK signal is the scan clock used for shift and capture operations of the boundary scan chain. It is one bit wider per channel.

The boundary scan registers are utilized in the JESD-229 standard during both functional and test modes. In functional mode, SSEN is low, CS_n is determined as needed to access appropriate memory blocks, and the other control signals do not effect operation. Beside functional mode, the JESD-229 standard supports five separate test modes as follow:

1. *Serial In/Out*—This mode is utilized to shift data through the boundary scan registers. During this operation, SSEN is high, SCK is running, and all other control signals are low.
2. *Serial In (No Out)*—This mode allows for only scan-in operations, as the SDO driver is disabled. To enter this mode, SSEN and SOE_n are high, SCK is running, and all other control signals are low.
3. *Parallel In*—In this mode, the boundary scan registers capture data from their associated microbumps in the array. SCK is running and all other control signals are high in this mode.
4. *Parallel Out*—In this mode, the boundary scan registers drive the output of their associated microbumps. In this mode, CS_n is low, all other control signals are high, and SCK has no effect on the operation.
5. *No-Operation (NOP)*—This mode disables both the data and SDO drivers, effectively preventing the boundary registers from changing state. In this mode, SSH_n is low, SSEN and CS_n are high, and no other control signals or the scan clock effect the circuit.

The P1838 die-level test wrapper discussed in Sect. 7.2 can be extended to utilize the boundary scan capabilities of the JESD-229 standard to perform post-bond interconnect testing between the logic die and the memory dies stacked on top of it. DRAM control signals are generated in the die wrapper on the logic die, and the DRAM boundary scan registers are included in the serial and parallel TAMs of the die wrapper. The wrapper extension supports up to four memory dies, just as the JEDEC standard does.

During interconnect test, the interconnects between the logic die and a chosen memory block are tested. The algorithmic test approach developed in [108] for the extension of the P1838 die wrapper tests interconnects rank by rank. Other approaches are also possible, for example testing memory blocks that belong to different channels on multiple ranks simultaneously.

Figure 7.13 shows the extension of the P1838 1500-style die wrapper to include interconnect testing of the DRAM stack. Although only two ranks and one channel are shown, the architecture is the same for all ranks and channels. An 1149.1 variant of the die wrapper will not be developed in this section, although the implementation is similar.

In the wrapper extension, the WSC signals are broadcast and, along with the WIR chain, handle all test control. The WIR is extended by 13 bits to accommodate a number of additional instructions. These instructions are used for enabling the DRAM boundary scan modes and selecting the blocks, channels, and ranks to test at any given moment. Several extra signals are generated by the WIR as follows:

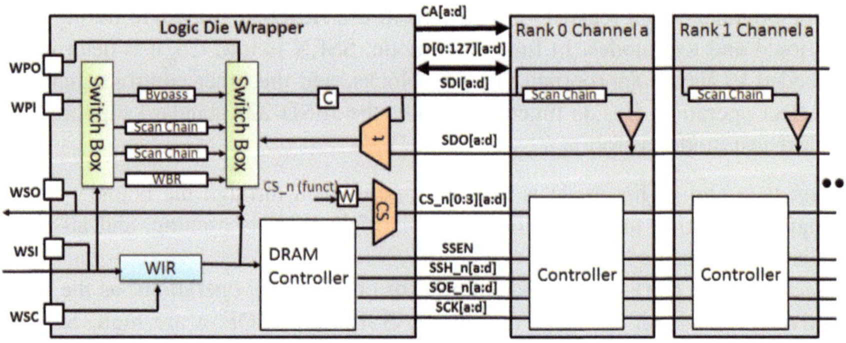

Fig. 7.13 Extension of the P1838 1500-style die wrapper for controlling test of the Wide-I/O DRAM

1. *DRAM_TE*—This one-bit *DRAM test enable* signal, when high, places the DRAM stack into test mode. Only a single bit is needed for the entire stack, as all ranks are set in test mode simultaneously. When low, the memories are in functional mode. The DRAM_TE signal will be equivalent in value to the SSEN signal that is internal to each rank. When the 3D wrapper is reset, DRAM_TE is set low.
2. *DRAM_ELEVATE*[a:d]—This signal contains one bit per DRAM channel. The signal controls the multiplexer marked t in Fig. 7.13. When a bit is high, the corresponding channel is placed in an elevate mode and is included in the TAM configuration of the stack. When a bit is low, the corresponding channel is in a turn mode and is bypassed by the TAM. For example, the bit sequence {1110} elevates test data to channels a, b, and c in the stack, but a turn is made at channel d and no test data enters it.
3. *DRAM_RS*[a:d]—This signal includes two pits per DRAM channel. This *DRAM rank select* signal selects one of the four ranks in a given channel. For example, the bit sequence 11 00 10 01 selects rank 3 in channel a, rank 0 in channel b, rank 2 in channel c, and rank 1 in channel d.
4. *DRAM_CAP*—This signal is one bit wide per memory channel. It sets test data to be captured in the memory die when high, or to be captured in the logic die when low.

Depending on the instruction in the WIR, the DRAM controller on the logic die generates the signals CS_n, SSEN, SSH_n, SOE_n, and SCK utilized by the JESD-229 boundary test logic. The test modes of the JESD-229 boundary scan are included in the P1838 wrapper modes as follows. The P1838 serial in and out shift modes activate the SDO drivers in the ranks determined by DRAM_RS. In this mode, test data is shifted in and out through SDI and SDO, respectively. The P1838 parallel in mode disables the tri-state drivers on the memory dies. The WBR on the logic die is then used to drive the memory data pins. The values on the WBR are then captured in the boundary scan registers on the memory dies, testing the drivers of the logic die and receivers of the memory dies. The P1838 parallel out mode selects the data drivers for the chosen ranks to drive their interconnects. The values driven in the memory dies are captured in the WBR of the logic die. This tests the drivers of the memory dies and the receivers of the logic die.

As can be seen in Fig. 7.13, a multiplexer labeled CS is included to switch between the functional CS_n signal and signals generated by the DRAM controller. Because the CS_n signal toggles the tri-state drivers of the memory inputs and outputs and is generated by the memory controller, an alternative method of generating these signals must be included for test purposes. Otherwise, there may be conflicting signals on the signal bus. The CS multiplexer is used to switch between the functional CS_n signals generated by the memory controller and test signals generated by the DRAM controller in the wrapper. The control signal for the CS multiplexer comes from the DRAM_TE signal.

The drawback of disabling the functional CS_n signal during test is that those signal paths are not tested during test of the internal die logic on the die wrapper. To compensate for this, scan registers that latch the functional CS_n values are inserted before the CS multiplexer. This provides the observability that is needed to test the signal paths. These scan registers must also have some associated logic for disabling the tri-state DRAM drivers when the memory stack is inactive.

7.3.2 Summary

Section Summary

- The P1838 die wrapper is extended for testing the interconnects between logic and memory dies in the JESD-229 standard that includes boundary scan registers that are not compliant with the IEEE 1149.1 or 1500 standards.
- Additional signals are added to the WIR, which is extended by 13 bits, for creating the signals needed by the internal controllers of each DRAM rank.
- Observe-only scan registers are added to the functional CS_n signals for observability during internal test of the logic die.

7.4 Conclusions

This chapter discussed the emerging P1838 die-level wrapper for 3D stack testing and briefly examined the currently available JEDEC standard for stackable Wide-I/O DRAMS. The test features of the JESD-229 standard were then explored, and the P1838 wrapper was extended to perform interconnect test in JESD-229 compatible memory-on-logic stacks.

The P1838 die-level wrapper extends the IEEE 1500 and IEEE 1149.1 standards to create die-level wrappers for dies in a 3D stack. These die-level wrappers present a standard test interface that allows for easy integration into a stack and modular testing of dies. The wrapper supports a low-bandwidth pre-bond test mode through the deposition of oversized probe pads and a switch box for setting the wrapper into a pre-bond test mode. Full bandwidth serial and parallel tests are available for the post-bond stack and can be used for both partial and complete stack testing. Furthermore, the wrapper allows for 1149.1-compliant board-level test via the lowest or highest die in the stack. Area estimates show that the area overhead of the P1838 wrapper is negligible for industrial dies.

The extension of the P1838 wrapper to JESD-229-compliant memory-on-logic stacks leverages the boundary scan capabilities present in the JEDEC standard to enable post-bond interconnect test between the logic and memory dies. The WIR is extended by 13 bits to provide instructions to a new DRAM controller present in the wrapper. The DRAM controller interprets the WIR instructions to provide signals to the test logic on each DRAM rank. The DRAM tests are then subsumed into available P1838 test modes. Additional multiplexers and scan cells are added to remove contention on and add observability to portions of the memory bus.

Chapter 8
Test-Architecture Optimization and Test Scheduling

8.1 Introduction

Previous chapters have discussed issues associated with pre-bond KGD test and test standards for pre-bond and post-bond test. Post-bond testing is in many ways a less complex issue than pre-bond test, because external test pins can be utilized for test access and TSVs in bonded dies can be treated as interconnects during test (although testing the TSVs themselves may require testing for failure modes unique to TSVs and neighboring active devices). Nevertheless, new constraints for post-bond testing of a 3D stack must be considered, such as limited test access to dies depending on their position in the stack, multiple post-bond test insertions, and limitations on the addition of test TSVs between dies. Just as for pre-bond test, optimizations are needed to design 3D test architectures and test schedules to minimize the cost of post-bond test in order to achieve cost-effective KGD, partial stack, and known-good-stack (KGS) test throughout the manufacture flow of a product.

Memories are easier to stack compared to logic due to high yields after repair and simplified testing and design [10], and as such 3D memory stacks have already been manufactured [10]. Stacks that include memory stacked on logic [46] or multiple logic dies [47] are likely to be seen in the near future. Although 3D design-and-test automation is not yet fully mature for commercial exploitation, it is well on its way [11] and many commercial design tools are beginning to support varying degrees of 3D design. These tools need to be able to exploit the benefits of 3D technologies while taking into account the various design-related trade-offs. For example, in a TSV-based 3D-SIC, the number of TSVs available for test access is limited because of their associated chip area costs. Most TSVs are likely to be dedicated to functional access, power/ground, and clock routing.

Post-bond testing of core-based dies in 3D-SICs brings forward new challenges [42, 43]. In order to test the dies and associated cores, a Test Access Mechanism (TAM) must be included on the dies to transport test data to the cores, and a 3D TAM is needed to transfer test data to the dies from the stack input/output pins. TAM design in 3D-SICs involves additional challenges compared to TAM

B. Noia and K. Chakrabarty, *Design-for-Test and Test Optimization Techniques for TSV-based 3D Stacked ICs*, DOI 10.1007/978-3-319-02378-6_8,

design for 2D SOCs. In a 3D-SIC, a test architecture must be able to support testing of individual dies as well as testing of partial and complete stacks, and it is for this reason that the test standards discussed in Chap. 7 are being developed. These standards require compatible test architecture optimizations that must not only minimize the test length, but also minimize the number of TSVs used to route the 3D TAM, as each TSV has area costs associated with it and is a potential source of defects in a 3D-SIC. The test length is therefore dependent on the test architecture and test schedule and is constrained by a limit on the available test resources.

In this chapter, test architecture optimization for 3D stacked ICs implemented using TSVs is discussed. The optimizations are compatible with emerging standards presented in Chap. 7. A variety of design cases for 3D SICs with die-level test architectures are considered—including dies with fixed test architectures and dies whose test architectures have yet to be designed. Over the course of this chapter, mathematical programming techniques are derived to create optimal solutions for a variety of architecture optimization problems. These mathematical models, in addition to being immediately useful for optimization, provide a clear framework for 3D optimization that serve as a foundation for future research and applications. The optimization will first be developed for a complete stack test, and then extended to include optimizations for any or all post-bond test insertions as well as post-bond TSV test. From this, it will be demonstrated that optimal test architecture solutions and test schedules for multiple test insertions are different from their counterparts for a final stack test alone.

8.1.1 3D Test Architecture and Test Scheduling

The problem of test-architecture optimization for 3D-SICs considers three different 3D integration cases—(1) *hard dies*, in which a test architecture already exists, (2) *soft dies*, for which the 2D (per die) and 3D (per stack) test architectures are co-optimized, and (3) *firm dies*, in which a test architecture already exists but serial/parallel conversion hardware may be added to the die in order to reduce test pin and TSV use and achieve better test resource allocation for stack testing. For the sake of simplicity and ease of implementation, this chapter assumes session-based test scheduling [15], i.e., in which all tests which are executed simultaneously need to be completed before the next test session is started. Methods for minimizing the number of TSVs or test pins used for target test lengths are developed for both a total stack limit and a limit on TSVs between neighboring dies. While it is theoretically possible to have multiple dies on a given layer in a stack, for this chapter it is assumed that there is only one die per layer in a stack. Furthermore, a core is considered to be part of a single die only, i.e., “3D cores” are not considered. In addition to minimizing the test length for each soft die, the test length for the complete stack is minimized in all three problem instances.

Testing of 2D SOCs and the optimization of related test-access architectures have been well studied [22, 24, 30, 32]. Optimization methods have included integer linear programming (ILP) [22], rectangle packing [22,31], iterative refinement [32], and other heuristics [24,33]. However, these methods were all originally developed for 2D SOCs, and the added test complexities related to 3D technology were not considered.

Recently, early work has been reported on testing of 3D-SICs. Heuristic methods for designing core wrappers in 3D-SICs were developed in [12]. These methods do not address the problem of 3D TAM design. ILP models for test architecture design for each die in a stack are presented in [48]. While these ILP models take into account some of the constraints related to 3D-SIC testing such as a TSV limit, this approach does not consider the reuse of die-level TAMs. A TAM wire-length minimization technique based on simulated annealing is presented in [49]. A drawback of this approach is that it implies a 3D test architecture that is not feasible in practice. Heuristic methods for reducing weighted test cost while taking into account the constraints on test pin widths in pre-bond and post-bond tests are described in [39]. An unrealistic assumption made in [39] is that TAMs can start and terminate in any layer.

In most prior work on 3D-SIC testing, TAM optimization is performed at die-level only, which leads to inefficient TAMs and non-optimal test schedules for partial-stack and complete-stack tests. Furthermore, all previous methods assume that the designer can create TAM architectures on each die during optimization, which may not be possible in all cases. In Chap. 7 a die-level wrapper and associated 3D architecture is presented to allow for all pre-bond and post-bond tests. This approach proposes die-level wrappers and leverages the current IEEE 1149.1 and IEEE 1500 standards. In addition to functional and test modes, die-level wrappers allow bypass of test data to and from higher die in the stack and reduced test bandwidth during pre-bond tests. This is a realistic and practical look at test architectures in 3D-SICs, but it offers no insight into optimization and test scheduling. The optimization methods presented in this chapter are compatible with die wrappers, and they do not make any unrealistic assumptions about die wrappers or the 3D TAM. Pre-bond testing is not included in the optimization in this chapter. If reconfigurable scan chains as described in Chap. 5 are utilized, then pre-bond test configurations for each die can be considered as a separate optimization problem.

8.1.2 The Need for Optimization Considering Multiple Post-Bond Test Insertions and TSV Test

Compared to two-dimensional ICs that typically require two test insertions, namely wafer test and package test, 3D stacking introduces a number of natural test insertions [42]. Because the die-stacking steps of thinning, alignment, and bonding can introduce defects, there may be a need to test multiple subsequent (partial) stacks during assembly. Figure 8.1 shows an example manufacturing and test flow for a 3D stack. First, wafer test (i.e., pre-bond test) can be used to test die prior

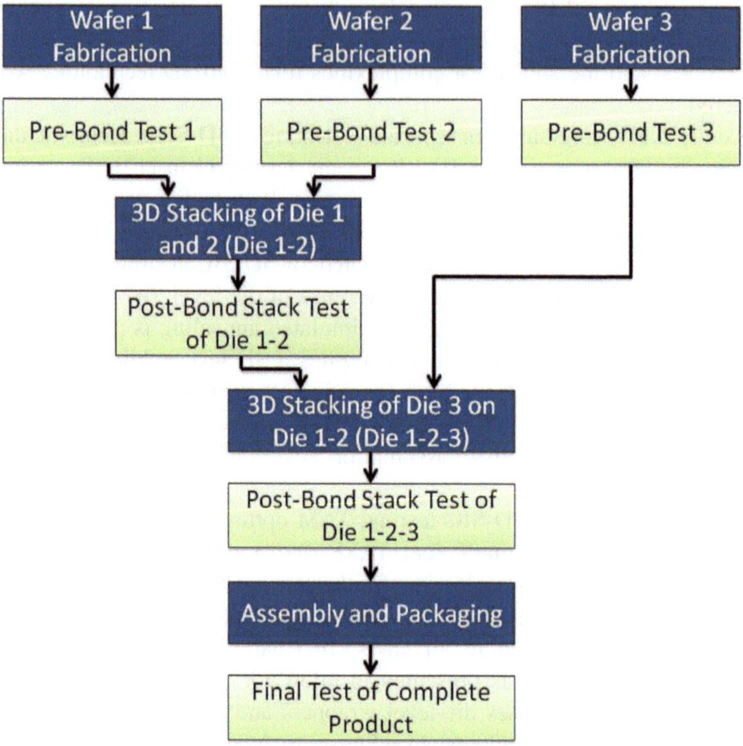

Fig. 8.1 3D-SIC manufacturing and test flow with multiple test insertions

to stacking to ensure correct functionality, as well as to match die in a stack for power and performance. Next, Die 1 and Die 2 are stacked, and then tested again. This is likely to be the first time the TSVs between Die 1 and Die 2 will be tested due to technology limitations that make pre-bond test of TSVs infeasible [43]. This step also ensures that defects can be detected in the stack due to additional 3D manufacturing steps such as alignment and bonding. The third die is added to the stack and all dies in the stack, including all TSV connections, are retested. Finally, the "known good stack" is packaged and the final product is tested. Optimization methods are needed to minimize test time not only for the final stack test, i.e., if the intermediate (partial) stacks are not tested, but also to minimize the total test time if the final stack and partial stacks are tested during bonding.

In Sect. 8.3, previously discussed optimization methods for 3D-SICs with *hard dies* and *soft dies* will be extended to consider multiple post-bond test insertions. In addition to minimizing the test time for each soft die, the test time can be minimized by considering all possible stack tests and the complete stack, as well as die-external tests as well. These optimization methods allow for the efficient generation of multiple options for testing a 3D-SIC.

In [23], the authors presented an expanded wrapper architecture for 2D ICs using modified wrapper cells in which each wrapper cell can be connected to two TAMs. As opposed to the 1500-standard wrapper (referred to in the rest of this chapter as a "thin" or 1500-like wrapper), this expanded wrapper architecture, or "fat" wrapper, allows for core-external test (EXTEST) and core-internal test (INTEST) to be run in parallel. This chapter will consider both types of wrappers for EXTEST optimization; in particular, the use of fat wrappers in this chapter is a natural extension of die-level wrappers to allow for die-external tests (TSV tests) and die-internal tests in parallel.

The rest of the chapter is organized as follows. In Sect. 8.2, optimization techniques are introduced for minimizing the test time for final stack test. A global limit is set on the number of dedicated TSVs to be used for test access and constraints are imposed on test bandwidth due to a limited number of test pins on the lowest die in the stack. While this optimization provides a sufficient starting point for designing a 3D test architecture, it does not consider multiple test insertions for testing the partial stack. Furthermore, the test time for TSVs and die-external logic is ignored in the optimization framework. Section 8.3 extends the model of Sect. 8.2 to allow for multiple test schedules and optimization for any number of or all post-bond stack tests. The test-bandwidth constraints use a more realistic model for dedicated test TSVs by considering a maximum number of TSVs per die, as opposed to a global limit. Furthermore, the test time for die-internal and die-external tests using both fat and thin wrappers is considered in the optimization. Section 8.6 concludes the chapter.

8.2 Test Architecture and Scheduling Optimization for Final Stack Test

In a 3D-SIC, which currently consist of anywhere from two to eight dies [13], the lowest die is usually directly connected to chip I/O pins and therefore can be tested using test pins. To test the non-bottom dies in the stack, test data must enter through the test pins on the lowest die. Therefore, to test other dies in the stack, the test access mechanism (TAM) must be extended to all dies in the stack through the test pins at the lowest die. To transport test data up and down the stack, "TestElevators" [35] need to be included on each die except for the highest die in the stack [42]. The number of test pins and TestElevators as well as the number of TSVs used affect the total test length for the stack.

Consider an example 3D-SIC with three dies with given test access architectures as shown in Fig. 8.2. Suppose the test lengths for Die 1, Die 2, and Die 3 are 300, 800, and 600 clock cycles, respectively. The total number of available test pins at the bottom die is 100. Die 1 requires 40 test pins (TAM width of 20), and Dies 2 and 3 require 60 TestElevators and 40 TestElevators, respectively. The test length for each die is determined by its test architecture.

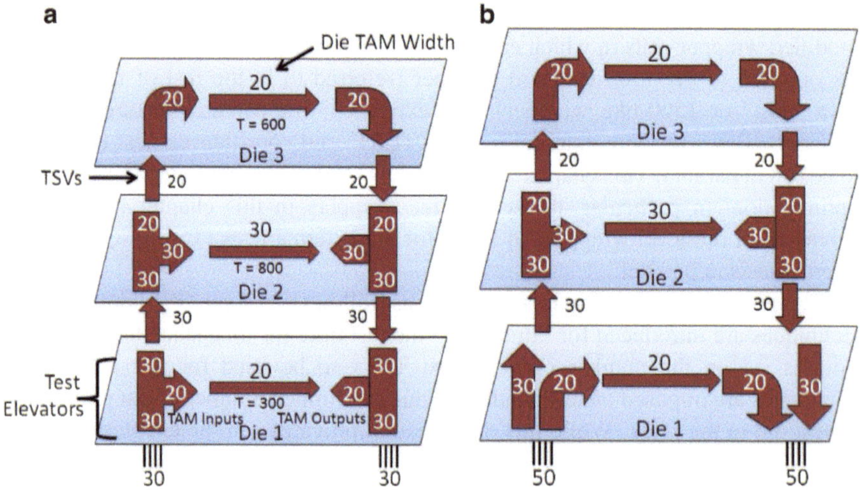

Fig. 8.2 Two example test architectures for a three-die stack with hard dies considering (**a**) serial die testing and (**b**) parallel testing of Die 1 and Die 2

Figure 8.2a shows the TestElevator widths and the number of TSVs used if all dies are tested serially. In this case, a total of 100 TSVs are used, and 100 test pins are available, of which only 60 are utilized. The total test length for the stack is the sum of the test lengths of the individual dies, i.e., 1700 cycles. Figure 8.2b shows the test architecture required if Die 1 and Die 2 are tested in parallel. In this case, the number of TSVs used is the same as in Fig. 8.2a. However, all 100 test pins are required to test Die 1 and Die 2 in parallel. Also, 60 TestElevators must pass between Die 1 and Die 2 in order to pass a separate 30-bit wide TAM to Die 2 for parallel testing. For this case, the total test length for the stack is $\max\{300, 800\} + 600 = 1400$ cycles. This example clearly shows that there is a trade-off between test length and the number of test pins and TSVs used. Therefore, a test-architecture optimization algorithm for 3D-SICs has to minimize the test length while taking into account upper limits on the number of test pins and TSVs used.

Test-architecture optimization for 3D-SICs with hard dies is illustrated in Fig. 8.3. For a hard die, the 2D test architecture on the die is fixed. The only structure over which the designer has control is the 3D TAM. Hard dies offer less flexibility for optimization in the sense that each die must have exactly the pre-defined number of input and output TAM wires appropriated to it in the design of the 3D TAM. Therefore, the only decisions that can be made in designing the 3D TAM is which (if any) dies can be tested in parallel with one another given the limitations on test pins and test TSVs. Hard dies may be present in TAM design problems if vendors sell fabricated dies to a 3D integrator.

Figure 8.3a illustrates the variables that arise for the hard-die problem. As can be seen, a fixed 2D TAM width is given along with the known test time for each die. The given constraints are the number of test pins W_{max} and the number of test TSVs TSV_{max} available. A solution, therefore, can be given as in Fig. 8.3b. Here, each die

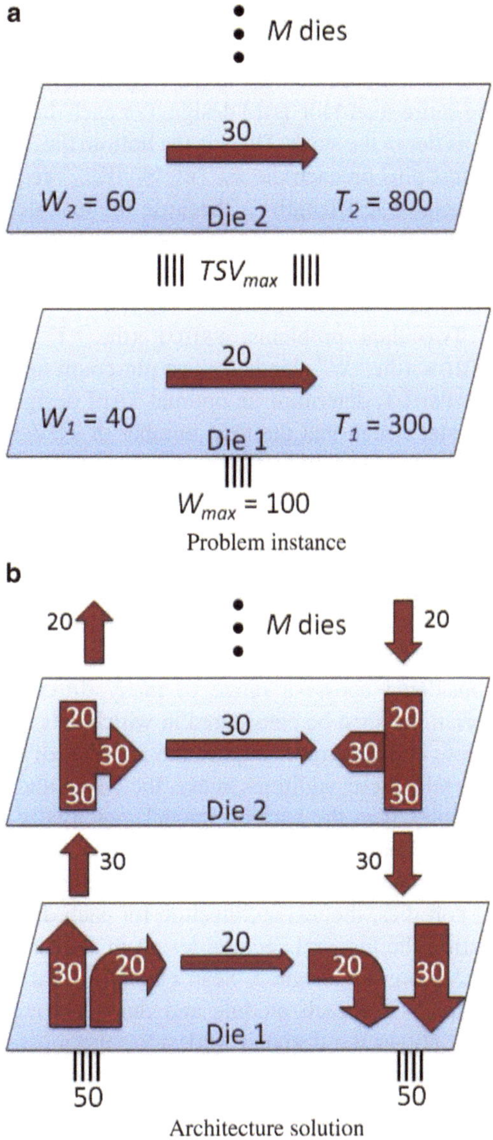

Fig. 8.3 Illustration of PSHD: (**a**) a problem instance; (**b**) an optimized architecture

receives the required and pre-defined test bandwidth, but Die 1 and Die 2 are tested in parallel through the 3D TAM.

The test-architecture optimization problem for hard dies is denoted as PSHD, where "PS" stands for "problem statement" and "HD" stands for "hard dies". The problem can be defined as follows.

3D-SIC with Hard Dies (PSHD)
Given a stack with a set M of dies, total number of test pins W_{max} available for test, and a maximum number of TSVs (TSV_{max}) that can be used globally (throughout the entire stack) for TAM design. For each die $m \in M$, the die's number corresponds to its tier in the stack (Die 1 is the bottom die, Die 2 is next, and so forth), the number of test pins on each die w_m ($w_m \leq W_{max}$) required to test the die is given, and the associated test length t_m (because the test architecture per die is given, t_m is also given). Determine an optimal TAM design and corresponding test schedule for the stack such that the total test length T for the stack is minimized and the number of TSVs used does not exceed TSV_{max}.

Two dual problems, PSHDT (the "T" stands for TSV minimization) and PSHDW (the "W" stands for test pin-count minimization), can be stated as follows. For PSHDT, determine an optimal TAM design and corresponding test schedule for the stack such that the total number of TSVs used for the stack is minimized and the upper limits on test length T_{max} and test pin count W_{max} are not exceeded. For PSHDW, determine an optimal TAM design and test schedule for the stack such that the total number of test pins used for the stack is minimized and the upper limits on test length T_{max} and total number of TSVs (TSV_{max}) are not exceeded.

The hard-die model is based on prior work on SOC testing [14] with additional constraints, while the firm and soft die models are considerably different and more complex. Besides simply adding 3D design constraints, each die must be considered across a range of many different possible TAM widths and many variations must be considered in which dies are tested in parallel along a 3D stack. These considerations require the addition of many more variables and constraints. Overall, these additions make the firm- and soft-die models significantly more complex than the hard-die model, potentially limiting the number of die that can be included in the model before run time becomes prohibitively high.

The above problem statement is different for a 3D-SIC with soft dies. In the case of soft dies, the test architecture for each die is not pre-defined, but is determined during the test-architecture design of the stack. In this case, both the 2D and 3D TAMs are co-designed. Scan chains for each test module are given, but the test wrappers for each module and the TAM are designed during 3D TAM design. This allows the designer to develop the most efficient 2D/3D TAM designs given TSV-count and test pin-count constraints. Soft dies model the additional flexibility available for optimization when dies are fabricated in-house for 3D integration.

Test-architecture optimization for 3D-SICs with soft dies is illustrated in Fig. 8.4. Figure 8.4a shows the known quantities associated with the soft die model, namely the number of modules per die, the pre-defined scan-chains per die, and W_{max} and TSV_{max}. Fig. 8.4b shows the result of optimization, including wrapper, 2D TAM, and 3D TAM designs.

The test-architecture optimization problem for soft dies can be formally defined as follows.

3D-SIC with Soft Dies (PSSD)
Given a stack with a set M of dies, the total number of test pins W_{max} available for test at the lowest die, and a maximum number of TSVs (TSV_{max}) that can be used for TAM design. For each die $m \in M$, the total number of cores c_m is given.

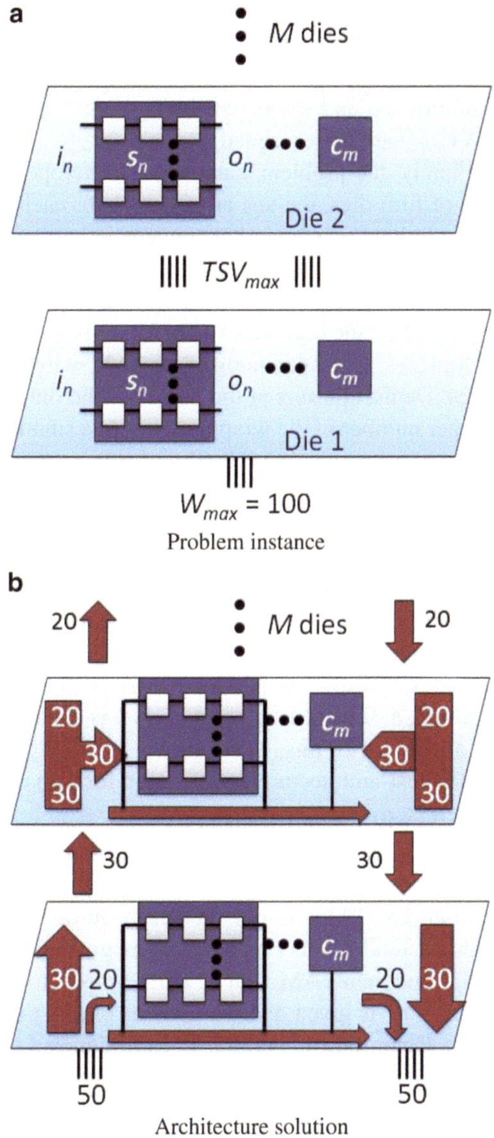

Fig. 8.4 Illustration of PSSD: (**a**) a problem instance; (**b**) optimized architecture

Furthermore, for each core c, the number of inputs i_c, outputs o_c, total number of test patterns p_c, total number of scan chains s_c, and for each scan chain v, the length of the scan chain in flip flops $l_{c,v}$ are given. Determine an optimal TAM design and test schedule for the stack, as well as for each die, such that the total test length T for the stack is minimized and the number of TSVs used does not exceed TSV_{max}.

Two dual problems, PSSDT and PSSDW, respectively, can again be stated as follows. For PSSDT, determine an optimal TAM design and test schedule for the stack and for each die such that the total number of TSVs used for the stack is minimized and the upper limits on test length T_{max} and test pin count W_{max} are not exceeded. For PSSDW, determine an optimal TAM design and test schedule for the stack as well as for each die such that the total number of test pins used for the stack is minimized and the upper limits on test length T_{max} and total number of TSVs (TSV_{max}) are not exceeded.

Finally, the problem statement is developed for a 3D-SIC with firm die. In the case of firm dies, the test architecture for each die is pre-defined as for a hard die, but additional serial/parallel conversion hardware may be added to the die to allow for fewer test elevators (or test pins in the case of the lowest die) to be used than in the case of the fixed 2D TAM width for the die. The conversion hardware is added before the inputs and after the outputs of the die wrapper. The input hardware multiplexes a smaller number of TAM wires to a larger number of die wrapper wires. Demultiplexers at the output of the die wrapper transfer test responses from a larger number of die wrapper wires to a smaller number of TAM wires. Compared to the scenario involving hard dies, this scenario allows the use of fewer test pins at the expense of higher test lengths, but also allows additional flexibility in test scheduling and test-time optimization.

The problem of test-architecture optimization for 3D-SICs with firm dies is shown in Fig. 8.5. Figure 8.5a shows the known quantities for the firm die problem; these are similar to those of the hard die problem except that test times are given for certain serial/parallel conversion bandwidths for each die. During optimization, one of these converters (or no converter) can be used, as seen in Fig. 8.5b. For Die 1, for example, a 3D TAM width of 15 bits is used, although the 2D TAM was designed for a width of 20 bits.

The test-architecture optimization for firm die is formally defined as follows.

3D-SIC with Firm Dies (PSFD)
Given a stack with a set M of dies, the total number of test pins W_{max} available for test at the lowest die, and a maximum number of TSVs (TSV_{max}) that can be used for TAM design. For each die $m \in M$, a fixed 2D TAM architecture with the total number of cores c_m is given along with which TAM partitions they utilize and their TAM widths. Furthermore, for each core n, the total number of test patterns p_n is given and the number of test pins $wlim_m$ required to test the die is given. Determine an optimal TAM design and test schedule for the stack, as well as possible serial/parallel conversion widths for each die, such that the total test length T for the stack is minimized and the number of TSVs used does not exceed TSV_{max}.

The above problems are all NP-hard ("proof by restriction"), as they can be reduced using standard techniques to the rectangle packing problem, which is known to be NP-hard [16]. For example, for PSHD, if the constraints on the maximum number of TSVs are removed, each die can be represented as a rectangle with a width equal to its test length and height equal to the number of required test pins. Now all these rectangles (dies) must be packed into a bin with a width equal to the total number of test pins and a height equal to the total test length for the stack, which must be minimized. Similarly, for PSSD, a rectangle must also be selected for each die from a set of rectangles with different widths and heights, but a special case of the scenario is identical to PSHD. Despite the NP-hard nature of these problems,

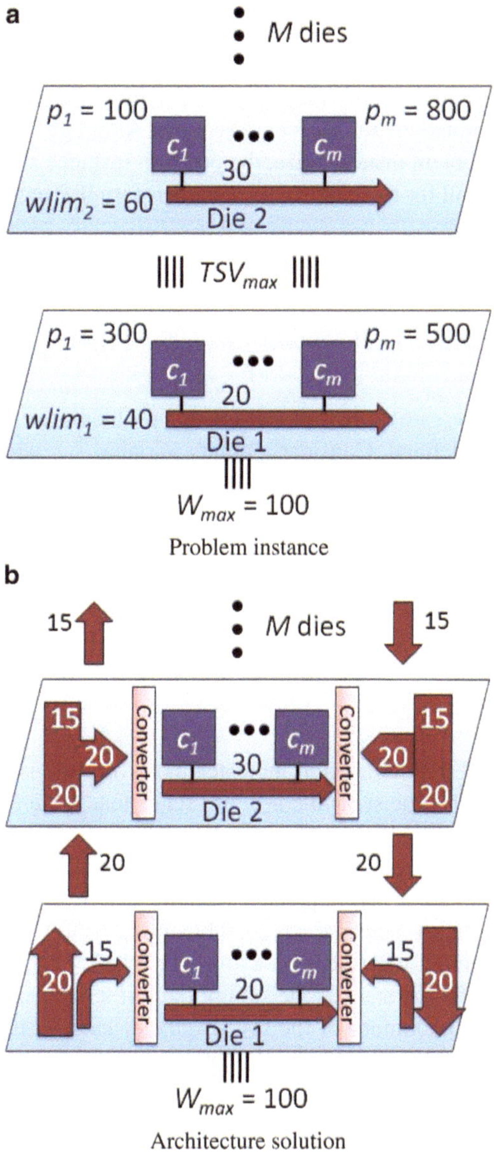

Fig. 8.5 Illustration of PSFD: (**a**) a problem instance; (**b**) optimized architecture

they can be solved optimally because the number of layers in a 3D-SIC is expected to be limited, e.g., up to four layers have been predicted for logic stacks [17].

The above problems are more general than the combinatorial problem of rectangle packing [16]. The added 3D design constraints and the greater design freedom available, especially for firm and soft dies, drastically increase the solution space. Rectangle packing is only a special and a considerably more simple case of our problem statements.

8.2.1 Test-Architecture Optimization for Final Stack Test

In this section, integer linear programming (ILP) is utilized to solve the problems defined in the previous section. Although ILP methods do not scale well with problem instance size, the problem instance sizes for P_{SHD} and P_{SSD} are relatively small for realistic stacks, and therefore ILP methods are good candidates for solving them.

8.2.2 ILP Formulation for PSHD

To create an ILP model for this problem, the set of variables and constraints must be defined. Consider a binary variable x_{ij}, which is equal to 1 if die i is tested in parallel with die j, and 0 otherwise. Constraints on variable x_{ij} can be defined as follows:

$$x_{ii} = 1 \qquad \forall i \tag{8.1}$$

$$x_{ij} = x_{ji} \qquad \forall i, j \tag{8.2}$$

$$1 - x_{ij} \geq x_{ik} - x_{jk} \geq x_{ij} - 1 \qquad \forall i \neq j \neq k \tag{8.3}$$

The first constraint indicates that every die is always considered to be tested with itself. The second constraint states that if die i is tested in parallel with die j, then die j is also tested in parallel with die i. The last constraint ensures that if die i is tested in parallel with die j, then it must also be tested in parallel with all other dies that are tested in parallel with die j.

Next, consider a second binary variable y_i, which is equal to 0 if die i is tested in parallel with die j on a lower layer ($l_i > l_j$), and 1 otherwise. The total test length T for the stack is the sum of the test lengths of all dies that are tested in series plus the maximum of the test lengths of each of the sets of parallel tested dies. Using variables x_{ij} and y_i, the total test length T for a stack with set of dies M can be defined as follows.

$$T = \sum_{i=1}^{|M|} y_i \cdot \left(\max_{j=i..|M|} \{x_{ij} \cdot t_j\} \right) \tag{8.4}$$

A proof of correctness of Equation 8.4 can be derived through induction as follows:

Base Case:

For the base case, consider two layers for which there are two possible optimization outcomes. Either both die are tested in series, or both die are tested in parallel. In the case of series testing, then $y_1 = 1, x_{11} = 1, x_{12} = 0$ and $y_2 = 1, x_{21} = 0, x_{22} = 1$. Using this information and equation 8.4, the test length is determined to be $y_1 \max\{x_{11} \cdot t_1\} + y_2 \max\{x_{22} \cdot t_2\} = \max\{t_1, 0\} + \max\{t_2\} = t_1 + t_2$. For parallel testing, the variables become $y_1 = 1, x_{11} = 1, x_{12} = 1$ and $y_2 = 0, x_{21} = 1, x_{22} = 1$. The equation becomes $1 \cdot \max\{t_1, t_2\} + 0 \cdot \max\{t_2\} = \max\{t_1, t_2\}$. These can both be demonstrated to be correct.

Induction Hypothesis:
It is assumed that (8.4) holds for M die.

Recursive Step:
It must be proven that the test length for die $M + 1$ is properly considered in the overall test length. Either die $M + 1$ is tested in serial with regard to the die in the stack, or it is tested in parallel with some die on a lower layer of the stack. When die $M + 1$ is tested in series, $y_{M+1} = 1, x_{M+1,M+1} = 1$, and $x_{n,M+1}$ and $x_{M+1,n}$ are zero for all $n \neq M + 1$. The test length becomes

$$\begin{aligned} &y_1 \cdot \max\{x_{11}t_1, x_{12}t_2, \cdots, x_{1M}t_M, x_{1,M+1}t_{M+1}\} + \\ &y_2 \cdot \max\{x_{22}t_2, x_{23}t_3, \cdots, x_{2M}t_M, x_{2,M+1}t_{M+1}\} + \\ &\qquad \cdots + y_M \max\{x_{MM}t_M\} \end{aligned} \tag{8.5}$$

In this equation, $x_{n,M+1}$ is 0 for all $n \neq M + 1$, so the test length of die $M + 1$ is only added to the total test length once, for:

$$y_{M+1} \max\{x_{M+1,M+1}t_{M+1} = t_{M+1}\} \tag{8.6}$$

For the parallel case, die $M + 1$ is tested in parallel with one or more die below it. Let die k be the die lowest in layer that is tested in parallel with die $M + 1$. Then $y_k = 1, x_{k,M+1} = 1, y_{M+1} = 0$, and $x_{M+1,k} = 1$. Then

$$y_{M+1} \max\{x_{M+1,M+1}t_{M+1}\} \tag{8.7}$$

goes to zero, and

$$y_k \max\{x_{kk}t_k, x_{k,k+1}t_{k+1}, \ldots, x_{k,M+1}t_{M+1}\} \tag{8.8}$$

takes into account the test length of dies k, $M + 1$, and any other die tested in this parallel tested set.

It should be noted that Eq. (8.4) has two non-linear elements, the *max* function, and the product of variable y_i and the *max* function. This equation is linearized by introducing two new variables. The variable c_i takes the value of the *max* function for each die i and the variable u_i represents the product $y_i \cdot c_i$. The variables u_i and c_i are defined using standard linearization techniques as shown in Fig. 8.6. The linearized function for total test length can be written as follows.

$$T = \sum_{i=1}^{|M|} u_i \tag{8.9}$$

As the number of test pins used for parallel testing of dies should not exceed the given test pins W_{max}, a constraint on the total number of pins used to test all dies in a parallel set can be defined as follows. In the inequalities, w_j refers to the TAM width for die j.

$$\sum_{j=1}^{|M|} x_{ij} \cdot w_j \leq W_{max} \qquad \forall i \tag{8.10}$$

Similarly, the total number of used TSVs should not exceed the given TSV limit TSV_{max}. The number of TSVs used to connect layer i to layer $i-1$ is the maximum of the number of pins required by the layer at or above layer i that takes the most test pin connections, and the sum of parallel-tested dies at or above layer i in the same parallel-tested set. Based on this, the constraint on the total number of TSVs used in a test architecture can be defined as follows:

$$\sum_{i=2}^{|M|} \{\max_{k=i}^{|M|}\{w_k, \sum_{j=k}^{|M|} w_j \cdot x_{kj}\}\} \leq TSV_{max} \tag{8.11}$$

The above set of constraints can be linearized by representing the max function by a variable d_i. Finally, to complete the ILP model for PSHD, constraints on binary variable y_i and the relationship between binary variable y_i and x_{ij} must be defined. For this purpose, a constant C is defined that approaches but is less than 1. It is then possible to define y_i as follows:

$$y_1 = 1 \tag{8.12}$$

$$y_i \geq \frac{1}{1-i} \sum_{j=1}^{i-1} (x_{ij} - 1) - C \qquad \forall i > 1 \tag{8.13}$$

Equation 8.12 forces y_1 to 1, because the lowest layer cannot be tested in parallel with any layer lower than itself. Constraint (8.13) defines y_i for the other layers. To understand this constraint, first make the observation that the objective function (as shown in Eq. (8.4)) would be minimized if each y_i is zero. This would make the objective function value equal to 0, which is an absolute minimum test length. Thus, y_i must be restricted to 1 only where it is absolutely necessary, otherwise the objective function can be relied on to assign a value 0 to all unrestricted y_i variables. This equation considers the range of values that the sum of x_{ij} can take. The fraction in the equation normalizes the sum to a value between 0 and 1 inclusive, while the summation considers all possible cases for a die being tested in parallel with a die below it.

Equation 8.13 can be proven correct through induction as follows:

Base Case:

Consider for the base case a stack of 2 die. The variable y_1 is always equal to one, $x_{11} = 1$, and x_{22}=1. There are two possible configurations for testing the two die. The first configuration is that both die are tested serially. If this is true, then the variables take the values of $x_{12} = 0$ and $x_{21} = 0$. The equation for y_2 then becomes:

$$y_2 \geq \frac{1}{1-2}(x_{21} - 1) - M \tag{8.14}$$

$$y_2 \geq 1 - M \tag{8.15}$$

Because $M < 1$, $1 - M$ is some small fraction greater than zero. Thus, y_2 must be greater than zero, and because it is binary it must take the value of one. The second possibility is that both die are tested in parallel such that $x_{12} = 1$ and $x_{21} = 1$. This defines y_2 as follows:

$$y_2 \geq \frac{1}{1-2}(x_{21} - 1) - M \tag{8.16}$$

$$y_2 \geq -M \tag{8.17}$$

This leaves y_2 unrestricted, as it can only take the value zero or one and is thus always greater than a negative number. Due to the objective, y_2 will become zero as desired, because it is tested in parallel with a die lower in the stack than itself.

Induction Hypothesis:

It is assumed that Eq. (8.13) holds for the case of m die.

Case of $m + 1$ die

For die $m + 1$, Equation (8.13) becomes:

$$\begin{aligned} y_{m+1} \geq \frac{1}{1-(m+1)}([x_{(m+1)1} - 1] + [x_{(m+1)2} - 1] + \\ \ldots + [x_{(m+1)m} - 1]) - M \end{aligned} \tag{8.18}$$

If $m + 1$ is tested serially, then the summation adds the quantity -1 a total of m times. This results in:

$$y_m \geq \frac{1}{-m}(-m) - M \tag{8.19}$$

$$y_m \geq 1 - M \tag{8.20}$$

This forces y_{m+1} to one. Now consider the range of values that can be taken by the right hand side of Eq. 8.18 for parallel testing cases of die $m + 1$ by considering the extremes. If die $m + 1$ is tested in parallel with only one die below it, then one of the terms of the summation becomes zero and the calculation becomes:

Fig. 8.6 ILP model for 3D TAM optimization PSHD

Objective:

Minimize $\sum_{i=1}^{|M|} u_i$

Subject to the Constraints:

$$
\begin{aligned}
&t_{max} = \max_{i=1}^{|M|} t_i \\
&c_i \geq x_{ij} \cdot t_j \qquad \forall i, j = i..|M| \\
&u_i \geq 0 \qquad \forall i \\
&u_i - t_{max} \cdot y_i \leq 0 \qquad \forall i \\
&u_i - c_i \leq 0 \qquad \forall i \\
&c_i - u_i + t_{max} \cdot y_i \leq t_{max} \qquad \forall i \\
&\sum_{j=1}^{|M|} x_{ij} \cdot w_j \leq W_{max} \qquad \forall i \\
&x_{ii} = 1 \qquad \forall i \\
&x_{ij} = x_{ji} \qquad \forall i, j \\
&1 - x_{ij} \geq x_{ik} - x_{jk} \geq x_{ij} - 1 \qquad \forall i \neq j \neq k \\
&\sum_{i=2}^{|M|} d_i \leq TSV_{max} \qquad \forall i \\
&d_i \geq \sum_{j=k}^{|M|} w_j \cdot x_{kj} \qquad \forall i, k = i..|M| \\
&d_i \geq w_j \qquad \forall i, j = i..|M| \\
&y_1 = 1 \\
&y_i \geq \frac{1}{1-i} \sum_{j=1}^{i-1} (x_{ij} - 1) - C \qquad \forall i > 1
\end{aligned}
$$

$$y_m \geq \frac{1}{-m}(-(m-1)) - M \tag{8.21}$$

$$y_m \geq \frac{1-m}{-m} - M \tag{8.22}$$

The fraction clearly results in a positive number less than one, and subtracting M makes this a negative value, leaving y_{m+1} unrestricted. In the case of die $m+1$ being tested in parallel with every die below it, every term of the summation is zero and the calculation becomes:

$$y_m \geq \frac{1}{-m}(0) - M \tag{8.23}$$

$$y_m \geq -M \tag{8.24}$$

Thus, for all cases of die $m+1$ tested in parallel, the right hand side is in the range $[\frac{1-m}{-m} - M, -M]$, which are all negative values, thereby leaving y_{m+1} unrestricted.

The complete ILP model for problem PSHD is shown in Fig. 8.6.

A benefit of using ILP is that the dual problems PSHDT and PSHDW can be easily tackled by appropriately modifying the model of Fig. 8.6. For both PSHDT and PSHDW, a maximum test length constraint T_{max} is introduced and the following inequality is added to the model:

$$\sum_{i=1}^{|M|} u_i \leq T_{max}.$$

As can be easily seen, the previous objective function is now transformed into a constraint. For PSHDT, the constraint on TSVs used is removed, which is the inequality involving TSV_{max}, and is replaced with the following objective function:

$$\text{Minimize} \sum_{i=2}^{|M|} d_i$$

For PSHDW, the constraint on the number of test pins used is removed, which is the inequality involving W_{max}, and the variable P is introduced to represent the number of test pins used by the stack. The following inequalities define P.

$$P \geq \sum_{j=1}^{|M|} x_{ij} \cdot w_j \qquad \forall i \tag{8.25}$$

Our objective for PSHDW is therefore to minimize P.

8.2.3 *ILP Formulation for* PSSD

The ILP formulation for 3D-SICs with soft cores is derived in a similar manner as that for 3D-SICs with hard cores. In this case, the test length t_i for die i is a function of the TAM width w_i assigned to it. Using the variables x_{ij} and y_i as defined in Sect. 8.2.2, the total test length T for the stack with the set of soft dies M can be defined as follows.

$$T = \sum_{i=1}^{|M|} y_i \cdot \max_{j=i..|M|} \{x_{ij} \cdot t_j(w_j)\} \tag{8.26}$$

It should be noted that Eq. (8.26) has several non-linear elements. To linearize this equation, the test length function must first be defined. For this purpose, the binary variable g_{in} is introduced where $g_{in} = 1$ if $w_i = n$, and 0 otherwise. The expression is then linearized using the variable v_{ij} for $x_{ij} \cdot \sum_{n=1}^{k_i}(g_{jn} \cdot t_j(n))$. Similarly to Eq. (8.9), the variable c_i takes the value of the max function for each die i and the variable u_i represents the product $y_i \cdot c_i$. Because w_j is now a decision variable, $x_{ij} \cdot w_j$ is linearized using a new variable z_{ijk} defined for all i, j, k. The max function is represented by the variable d_i as before. By using the variable z_{ijk}, the TAM width that can be given to each die can be constrained by an upper limit, which is the number of available test pins. This is represented with the following set of inequalities. The complete ILP model for PSSD is shown in Fig. 8.7.

$$\sum_{j=1}^{|M|} z_{jij} \leq W_{max} \qquad \forall i \tag{8.27}$$

Fig. 8.7 ILP model for 3D TAM optimization PSSD

Objective:

Minimize $\sum_{i=1}^{N} u_i$

Subject to the Constraints:

$$
\begin{array}{ll}
t_{max} = \max_{i=1}^{|M|} t_i & \\
c_i \geq v_{ij} & \forall i, j = i..|M| \\
v_{ij} \geq 0 & \forall i, j \\
v_{ij} - t_{max} \cdot x_{ij} \leq 0 & \forall i, j = 1..|M| \\
-\sum_{n=1}^{k_i} (g_{jn} \cdot t_j(n)) + v_{ij} \leq 0 & \forall i, j \\
\sum_{n=1}^{k_i} (g_{jn} \cdot t_j(n)) - v_{ij} + t_{max} \cdot x_{ij} \leq t_{max} & \forall i, j \\
u_i \geq 0 & \forall i \\
u_i - t_{max} \cdot y_i \leq 0 & \forall i \\
u_i - c_i \leq 0 & \forall i \\
c_i - u_i + t_{max} \cdot y_i \leq t_{max} & \forall i \\
z_{ijk} \geq 0 & \forall i, j, k \\
z_{ijk} - t_{max} \cdot x_{jk} \leq 0 & \forall i, j, k \\
-wi + z_{ijk} \leq 0 & \forall i, j, k \\
w_i - z_{ijk} + t_{max} \cdot x_{jk} \leq t_{max} & \forall i, j, k \\
\sum_{i=2}^{|M|} d_i \leq TSV_{max} & \\
d_i \geq \sum_{j=k}^{|M|} z_{jkj} & \forall i, k = i..|M| \\
d_i \geq w_j & \forall i, j = i..|M| \\
\sum_{j=1}^{|M|} z_{jij} \leq W_{max} & \forall i \\
x_{ii} = 1 & \forall i \\
x_{ij} = x_{ji} & \forall i, j \\
1 - x_{ij} \geq x_{ik} - x_{jk} \geq x_{ij} - 1 & \forall i \neq j \neq k \\
y_1 = 1 & \\
y_i \geq \frac{1}{1-i} \sum_{j=1}^{i-1} (x_{ij} - 1) - M & \forall i > 1
\end{array}
$$

As before, alterations are made to the ILP model to solve the dual problems PSSDT and PSSDW. For both PSSDT and PSSDW, as with the hard die dual problems, a maximum test length constraint T_{max} is introduced and the following constraint is added to the problem:

$$\sum_{i=1}^{|M|} u_i \leq T_{max} \tag{8.28}$$

For PSSDT, the constraint on TSVs used is removed and the following objective function is used:

$$\text{Minimize} \sum_{i=2}^{|M|} d_i.$$

For PSHDT, the constraint on the number of test pins used is removed, which is the inequality involving W_{max}, and variable P is once again utilized. The following inequality defines P.

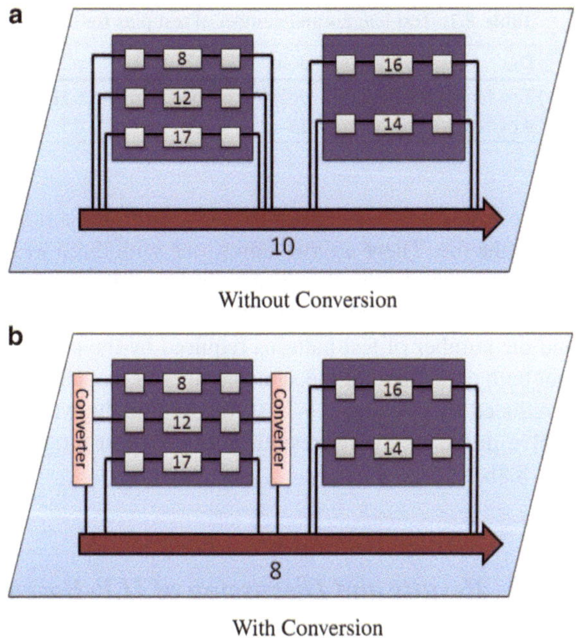

Fig. 8.8 Illustration of TAM width reduction using serial/parallel conversion (**a**) Without Conversion (**b**) With Conversion

$$P \geq \sum_{j=1}^{|M|} z_{jij} \qquad \forall i \tag{8.29}$$

Our objective for PSHDT is therefore to minimize P.

8.2.4 *ILP Formulation for* PSFD

The ILP formulation for 3D-SICs with firm dies is an extension of the model for soft dies. A constraint is added to indicate that the number of test pins used for a die cannot exceed the number of test pins required by the fixed 2D TAM for that die. This constraint is expressed as:

$$w_i \leq wlim_i \qquad \forall i \tag{8.30}$$

where $wlim_i$ is the number of test pins required by the 2D TAM on each die i prior to any serial/parallel conversion. In order to accurately determine test lengths for the dies using serial/parallel conversion, the control-aware TAM design method of [18] is modified to allow the architecture to be fixed in terms of assignment of modules to TAM partitions for a die. The effective widths of the TAM partitions are then iteratively reduced and re-optimized, thereby determining the optimal serial/parallel conversion to use depending on the bandwidth given to that die as shown in Fig. 8.8.

Table 8.1 Test lengths and number of test pins for dies as required in PSHD

Die	d695	f2126	p22810	p34392	p93791
Test length (cycles)	96297	669329	651281	1384949	1947063
# of Test pins	15	20	25	25	30

Figure 8.8a shows a die prior to TAM width reduction, where ten pins are required to test the die. There are two cores, one with three wrapper chains consisting of the given number of scan flops, and another with two wrapper chains. The amount of time needed to test each core is dependent on the length of the longest wrapper chain and the number of test patterns required by the core. In this example, it is assumed that both cores require the same number of test patterns. Therefore, the TAM width is reduced by two and it is best to combine the wrapper chains of length eight and twelve in the first core, resulting in a longest wrapper chain of twenty as seen in Fig. 8.8b.

8.2.5 Results and Discussion of ILP-Based Final Stack Test Optimization

In this section, experimental results are shown for the ILP models presented in the previous section. As benchmarks, three 3D-SICs (as shown in Fig. 8.9) have been handcrafted from several SOCs from the ITC'02 SOC Test Benchmarks as dies inside the 3D-SICs. The SOCs used are d695, f2126, p22810, p34292, and p93791. In SIC 1, the die are ordered such that the lowest die is the most complex (p93791), with dies increasing in complexity as one moves higher in the stack. The order is reversed in SIC 2, while for SIC 3, the most complex die is placed in the middle of the stack, with dies decreasing in complexity moving out from that die. For equal test bitwidths, the dies lowest in the stack in SIC 1 have the highest test times. In Table 8.1, f2126 has a slightly higher test time than p22810 because it has a smaller test bitwidth. P22810, however, is still the more complex die from a test perspective. Because SIC 1 and SIC 2 are two extreme cases, they better illustrate the results that are generated. SIC 3 is included to demonstrate test times for an intermediate case of 3D stacking, as opposed to simply the opposite extremes.

To determine the test architecture and test length for a given die (SOC) with a given TAM width, the control-aware TAM design method in [18] has been used. Control-aware TAM design takes into account the number of scan-enable signals required for independent testing of TAMs in the architecture. For PSHD (3D-SIC with hard dies), the test lengths (cycles) and TAM widths for different dies are listed in Table 8.1. Note that test pins were assigned to dies based on their sizes in order to avoid very large test lengths for any individual die.

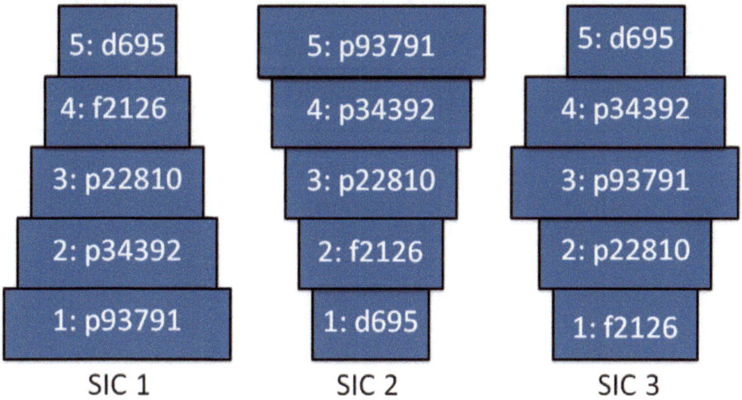

Fig. 8.9 Three 3D-SIC benchmarks

Table 8.2 Comparison of optimization results between PSHD and a greedy algorithm for SIC 1

		PSHD (ILP)		PSHD (Greedy)		Percentage difference
TSV_{max}	W_{pin}	Test length (cycles)	Test schedule	Test length (cycles)	Test schedule	in test length (ILP versus Greedy)
160	30	4748920	1,2,3,4,5	4748920	1,2,3,4,5	0.0
160	35	4652620	1,2,3,4‖5	4652620	1,2,3,4‖5	0.0
160	40	4652620	1,2,3,4‖5	4652620	1,2,3,4‖5	0.0
160	45	3983290	1‖5,2‖4,3	4001340	1,2‖5,3‖4	0.5
160	50	3428310	1‖4,2‖3,5	3428310	1‖4,2‖3,5	0.0
160	55	2712690	1‖2,3‖4,5	2712690	1‖2,3‖4,5	0.0
160	60	2616390	1‖2,3‖4‖5	2712690	1‖2,3‖4,5	3.7
160	65	2616390	1‖2,3‖4‖5	2712690	1‖2,3‖4,5	3.7
160	70	2616390	1‖2‖5,3‖4	2616390	1‖2‖5,3‖4	0.0
160	75	2598340	1‖2‖4,3‖5	2616390	1‖2‖5,3‖4	0.7
160	80	2598340	1‖2‖4,3‖5	2616390	1‖2‖5,3‖4	0.7
160	85	2598340	1‖2‖4,3‖5	2616390	1‖2‖5,3‖4	0.7
160	90	2598340	1‖2‖4,3‖5	2616390	1‖2‖5,3‖4	0.7
160	95	2043360	1‖2‖3‖4,5	2616390	1‖2‖5,3‖4	28.0
160	100	2043360	1‖2‖3‖4,5	2043360	1‖2‖3‖4,5	0.0

The minimal achievable test length for the hard die stack can be seen to be 1947063 cycles, which occurs when all dies are tested in parallel with one another. To investigate the effect of achieving this test length for a 3D stack, consider SIC 1 and SIC 2. For both SICs, this architecture requires 115 test pins on the bottom die. For SIC 1, this requires 195 test TSVs. For SIC 2, this requires 265 test TSVs.

Table 8.2 compares optimal results produced using ILP with those produced using a greedy algorithm for PSHD SIC 1. The greedy algorithm attempts to combine dies in parallel-tested sets, starting with those that would lead to the greatest reduction in test time. If more than one combination results in the same reduction, it

Table 8.3 Comparison of optimization results between PSSD and a greedy algorithm for SIC 1

		PSSD (ILP)		PSSD (Greedy)		Percentage difference
TSV_{max}	W_{pin}	Test length (cycles)	Test schedule	Test length (cycles)	Test schedule	in test length (ILP versus Greedy)
140	30	4795930	1‖2‖3‖4,5	7842000	1‖2,3,4‖5	63.5
140	35	4237100	1‖2‖3‖4,5	7633580	1‖3,2‖4,5	80.1
140	40	3841360	1‖2‖3‖4,5	6846400	1‖3,2‖4,5	78.2
140	45	3591550	1‖2‖3‖4,5	6379510	1‖2‖3,4‖5	77.6
140	50	3090720	1‖2‖3‖4,5	6041270	1‖2‖3,4‖5	95.5
140	55	2991860	1‖2‖3‖4‖5	5873430	1‖2‖3‖4,5	96.3
140	60	2873290	1‖2‖3‖4,5	5821900	1‖2‖3‖4,5	102.6
140	65	2784050	1‖2‖3‖4‖5	5705410	1‖2‖3‖4,5	104.9
140	70	2743320	1‖2‖3‖4‖5	5638140	1‖2‖3‖4,5	105.5
140	75	2629500	1‖2‖3‖4‖5	5638140	1‖2‖3‖4,5	114.4
140	80	2439380	1‖2‖3‖4‖5	5496200	1‖2‖3‖4,5	125.3
140	85	2402330	1‖2‖3‖4‖5	5447190	1‖2‖3‖4,5	126.7
140	90	2395760	1‖2‖3‖4‖5	5447190	1‖2‖3‖4,5	127.4
140	95	2383400	1‖2‖3‖4‖5	5447190	1‖2‖3‖4,5	128.5
140	100	2369680	1‖2‖3‖4‖5	5351480	1‖2‖3‖4,5	125.8

prioritizes those combinations that result in the smallest test resource use. Compared to PSSD and PSFD, PSHD is a less complex problem. Thus, the greedy algorithm is capable of producing results that are sometimes optimal, although it often does not result in the minimum test time.

Table 8.3 shows the information in Table 8.2 for problem PSSD. The soft die problem is more difficult to solve using a greedy heuristic. The heuristic algorithm for PSSD uses the greedy algorithm from PSHD as a subroutine in test-architecture optimization. It begins with an assignment of an equal number of test pins to each of the dies and optimizes the 2D and 3D TAM under these constraints. It then randomly adds and removes random numbers of test pins from each die, each time balancing the result to use the maximum number of test pins, and optimizes again. It checks for reductions in test time, returning to the best solution so far if no test time reduction is produced or constraints are violated. It terminates after 10000 iterations of no improvement. As can be seen, the optimal ILP solution tends to be much better in terms of test length than the heuristic solution, simply because the solution space is so large.

It is useful to briefly discuss how the test architecture for a soft die is built using the information produced from the ILP optimization. Prior to using the ILP model, 2D TAM architectures that minimize test time are produced for each die in the stack assuming a wide range of TAM widths available to that die. In this sense, the 2D architecture is already completed and a choice must be made regarding which architecture is to be used. The ILP formulation provides data regarding which TAM width to use for each die and the test schedule for testing the die, which lets the designer know which dies need to be tested in parallel. With this information, the design of the 3D TAM is elementary. The integrator simply provides the appropriate TAM width to each die, assuring that the number of test elevators between each die is sufficient for the bandwidth required for any parallel tests.

Take for example the information provided for the ILP optimization of soft dies in Table 8.3 when W_{pin} is 60. Although not shown in the table, the ILP optimization provides the designer with the following widths for each die (these values show the number of test pins used by that die, so the TAM width is half the given value): $W_1 = 30$, $W_2 = 20$, $W_3 = 6$, $W_4 = 4$, $W_5 = 30$. The designer sees that Dies 1-4 are tested in parallel, followed by Die 5. Because the width of the top die dominates this stack, TSV routing simply requires 30 test elevators between each die. Die 1, 2, 3, and 4 each utilize a different test pin of the 60 test pins available, and Die 5 can utilize any 30 of the test pins. Which pin is routed to which die is up to the best judgment of the designer, as is wire routing and the like.

For a fixed TSV_{max} and range of W_{max}, Table 8.4 presents representative results for PSHD for the three benchmark 3D-SICs using hybrid TestRail architectures [32]. Additional values for TSV_{max} could be considered, but they do not provide any new insights. For PSHD and its comparison to PSHD, optimizations were done using hybrid TestBus [32] architecture for variety. The ILP models were solved using the XPRESS-MP tool [34]. In this table, Column 1 shows the maximum number of TSVs allowed (TSV_{max}), while Column 2 represents the number of available test pins W_{max}. Columns 3, 6 and 9 represent the total test length (cycles) for the stack for 3D-SIC 1, 2 and 3 respectively. Columns 4, 7, and 10 show the resulting test schedule for the 3D-SICs, where the symbol "||" indicates parallel testing of dies, and a "," represents serial testing. Finally, Columns 5, 8, and 11 show the percent decrease in test length over the serial testing case for the three 3D-SICs. From Table 8.4 it can be seen that compared to serial testing of all dies (first row in the table), the proposed method obtains up to 57 % reduction in test length. Note that although identical test lengths were obtained for SIC 2 and SIC 3 for $TSV_{max} = 160$, different TAM architectures and test schedules were obtained from the optimization algorithm (see Columns 4 and 10).

For a different number of TSVs (TSV_{max}), Fig. 8.10a and b show the variation in test length T with an increase in number of test pins W_{max} for SIC 1 and SIC 2. From the figures, it can be seen that both TSV_{max} and W_{max} determine which dies should be tested in parallel, and thus determine the total test length for the stack. For a given value of TSV_{max}, increasing W_{max} does not always decrease the test length, showing the presence of pareto-optimal points. These have an important impact on test resource allocation, because test resources for a target test length should be provided only to the extent that they align with the first point in a pareto-optimal plateau.

Figure 8.11 shows the variation in test length for SIC 2 when both TSV_{max} and W_{max} are varied. From the figure, it can be seen that a small increase in the number of test pins W_{max} for a given TSV_{max} reduces test length significantly, while to achieve the same reduction in test length with a fixed number of test pins W_{max}, a large increase in TSV_{max} is required.

Table 8.4 Experimental results for PSHD

		PSHD SIC 1			PSHD SIC 2			PSHD SIC 3		
TSV_{max}	W_{pin}	Test length (cycles)	Test schedule	Reduction (%)	Test length (cycles)	Test schedule	Reduction (%)	Test length (cycles)	Test schedule	Reduction (%)
160	30	4748920	1,2,3,4,5	0.00	4748920	1,2,3,4,5	0.00	4748920	1,2,3,4,5	0.00
160	35	4652620	1,2,3,4‖5	2.03	4652620	1‖2,3,4,5	2.03	4652620	1‖5,2,3,4	2.03
160	40	4652620	1,2,3,4‖5	2.03	4652620	1‖3,2,4,5	2.03	4652620	1‖5,2,3,4	2.03
160	45	3983290	1‖5,2‖4,3	16.12	3983290	1‖3,2‖4,5	16.12	3983290	1‖4,2‖5,3	16.12
160	50	3428310	1‖4,2‖3,5	27.81	3428310	1,2‖5,3‖4	27.81	3428310	1,2‖4,3‖5	27.81
160	55	2712690	1‖2,3‖4,5	42.88	2712690	1,2‖3,4‖5	42.88	2712690	1,2‖3,4‖5	42.88
160	60	2616390	1‖2,3‖4‖5	44.91	2616390	1‖2‖3,4‖5	44.91	2616390	1‖4‖5,2‖3	44.91
160	65	2616390	1‖2,3‖4‖5	44.91	2616390	1‖2‖3,4‖5	44.91	2616390	1‖4‖5,2‖3	44.91
160	70	2616390	1‖2‖5,3‖4	44.91	2616390	1‖2‖3,4‖5	44.91	2616390	1‖2‖3,4‖5	44.91
160	75	2598340	1‖2‖4,3‖5	45.29	2616390	1‖2‖3,4‖5	44.91	2616390	1‖2‖3,4‖5	44.91
160	80	2598340	1‖2‖4,3‖5	45.29	2616390	1‖2‖3,4‖5	44.91	2616390	1‖2‖3,4‖5	44.91
160	85	2598340	1‖2‖4,3‖5	45.29	2616390	1‖2‖3,4‖5	44.91	2616390	1‖2‖3,4‖5	44.91
160	90	2598340	1‖2‖4,3‖5	45.29	2616390	1‖2‖3,4‖5	44.91	2616390	1‖2‖3,4‖5	44.91
160	95	2598340	1‖2‖4,3‖5	45.29	2616390	1‖2‖3,4‖5	44.91	2616390	1‖2‖3,4‖5	44.91
160	100	2043360	1‖2‖3‖4,5	56.97	2616390	1‖2‖3,4‖5	44.91	2616390	1‖2‖3,4‖5	44.91
160	105	2043360	1‖2‖3‖4,5	56.97	2616390	1‖2‖3,4‖5	44.91	2616390	1‖2‖3,4‖5	44.91

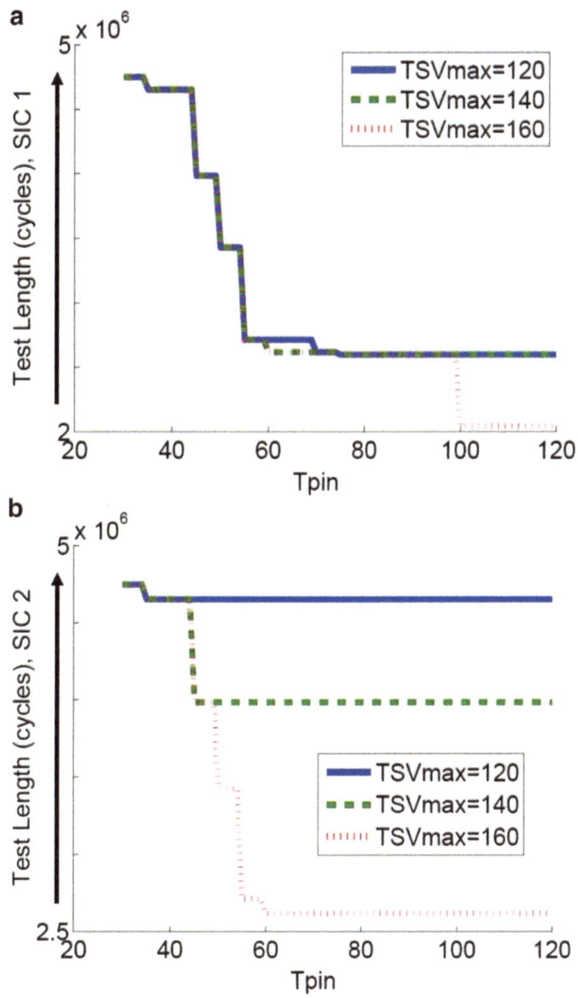

Fig. 8.10 The test length with respect to TSV_{max} for SIC 1 and SIC 2 with hard dies

Figure 8.12 demonstrates the differences between SIC 1 and SIC 2 during optimization. Two 3D stacks with five dies each are shown, with TAM widths displayed on each die and the number of TSVs used between each die shown to the left of the stack. Figure 8.12a shows the number of TSVs needed to test Die 1 and Die 2 in parallel followed by Dies 3, 4, and 5 for SIC 1. It is desirable to test Die 1 and Die 2 in parallel because they are the dies with the longest test lengths. This requires 90 TSVs. For SIC 2, this requires 250 TSVs as shown in Fig. 8.12b. This demonstrates why optimization produces better results for SIC 1 than for SIC 2.

Table 8.5 compares results for PSHD and PSFD for TestBus architectures. Table 8.5 demonstrates that by adding serial/parallel conversion of TAM inputs to hard dies, a reduction in test length as high as 28 % can be obtained. This is because the conversion allows for an increase in the test length of individual die in order to minimize the overall SIC test length during test schedule optimization. It should also

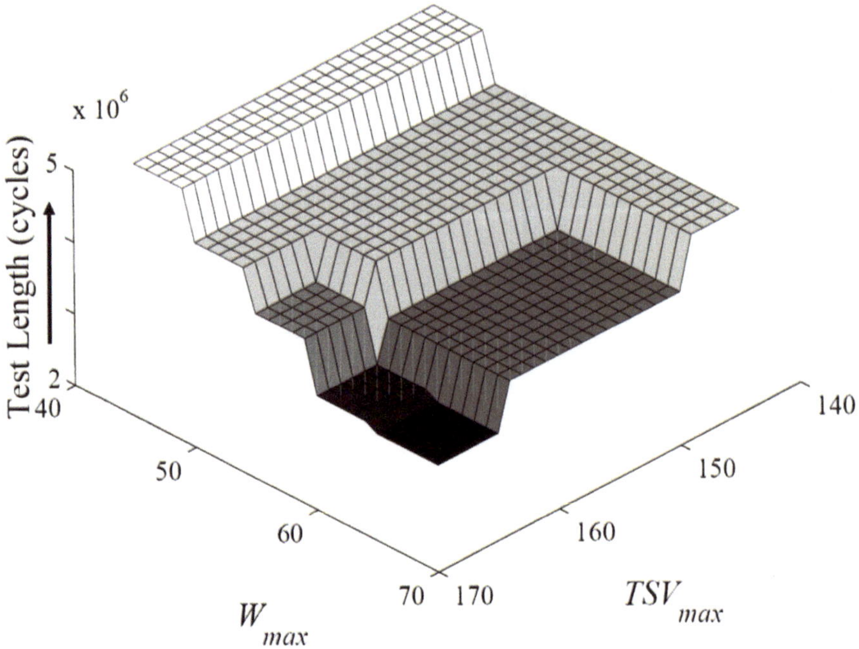

Fig. 8.11 Variation in test length with W_{max} and TSV_{max} for SIC 2 with hard dies

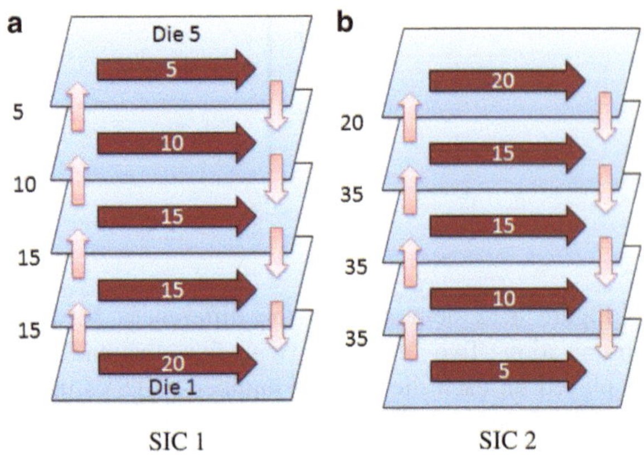

Fig. 8.12 Example of optimization for SIC 1 versus SIC 2 (**a**) SIC 1 (**b**) SIC 2

be noted that the test schedules and the number of test pins utilized for each die differ considerably between the hard-die and firm-die problem instances. Compared to a hard die, a firm die requires a small amount of extra hardware to convert a narrow TAM at the die input to a wider die-internal TAM, and vice versa. The area cost of this additional hardware is negligible compared to the die area and the hardware required for the core and die wrappers.

Table 8.5 Comparisons between PSHD and PSFD

Optimization framework	W_{max}	Test length (cycles)	Reduction (%)	Test schedule	No. of test pins used per Die
PSHD	35	4678670	0	1,2,3,4‖5	30,24,24,20,14
	44	4009340	0	1‖4,2,3‖5	30,24,24,20,14
	50	3381720	0	1‖3,2‖5,4,	30,24,24,20,14
	60	2658750	0	1‖5,2‖3,4,	30,24,24,20,14
	80	2658750	0	1‖5,2‖3,4,	30,24,24,20,14
PSFD	35	3828490	18.17	1‖4,2‖3,5	28,24,10,7,14
	44	2875900	28.27	1‖2,3‖4,5	28,16,24,18,14
	50	2641060	21.90	1‖2,3‖4‖5	30,18,24,18,4
	60	2335780	12.15	1‖2‖3‖4,5	28,16,10,6,14
	80	1971400	25.85	1‖2‖3‖4‖5	30,24,10,8,8

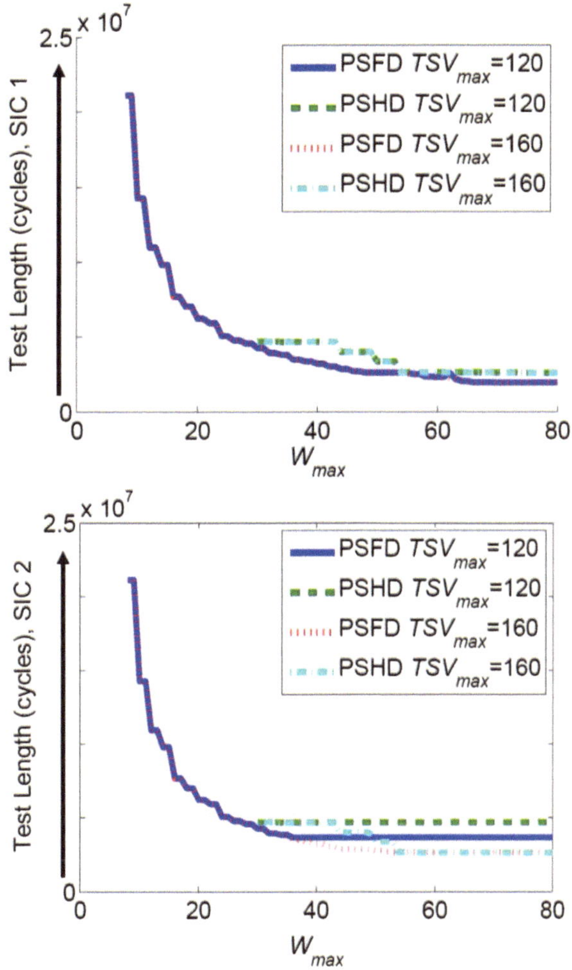

Fig. 8.13 Comparison of variation in test length with W_{max} for SIC 1 and SIC 2 between firm dies and hard dies

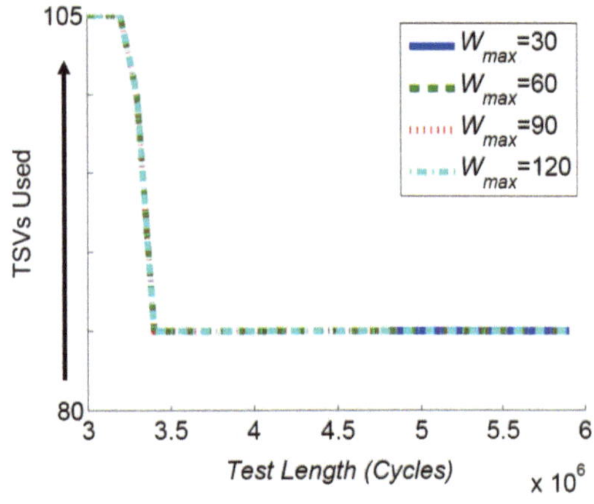

Fig. 8.14 Variation in TSVs used with T_{max} for SIC 1 with hard dies

Figure 8.13 shows comparative test lengths between PSHD and PSFD when W_{max} is varied, for two values of TSV_{max} and for two SICs. It is impossible to test the hard dies in these cases using fewer than 30 test pins without using serial/parallel conversion. As fewer test pins are used, the test lengths for individual dies greatly increase, resulting in a sharp increase in overall test length below certain values of W_{max}. It is important to note that the test length for a SIC with hard dies can never be shorter than the test length for the same SIC with firm dies; at best it can be equal. This is because, in the worst case with respect to test length, the optimization for firm dies is equivalent to the optimization for hard dies, i.e., no serial/parallel conversion is carried out. It can be seen that the use of serial/parallel conversion hardware can result in less use of test resources and shorter test time compared to hard dies without conversion. This observation is particularly valid in SIC 2, where the position of dies in the stack limits test time reduction (Fig. 8.16).

Figure 8.14 shows PSHDT (3D-SIC with hard dies and TSV-count optimization) results for SIC 1. Under tight test length constraints, a solution to the optimization is unattainable for smaller values of W_{max}. In Fig. 8.14, for example, a W_{max} value of 30 will not produce a feasible test architecture until the test length is above 470000 cycles. Once an optimized architecture is achievable, the minimum number of TSVs used is generally the same regardless of the value of W_{max}. There are two reasons for this. The first is that there are only a few configurations of the 3D TAM for the hard-die stack, and multiple configurations will give the same minimal TSV value. This is only a partial explanation, however, as equal minimal-TSV values for various W_{max} values are seen for soft dies as well. The primary reason for the results of Fig. 8.14 is that in order to minimize the number of TSVs used by the stack, the ILP solver will tend toward testing all dies in series with each other. If this cannot be done, then it will attempt to test only those dies with the smallest TAM width in parallel. This test configuration—tending toward serial testing—also happens to be the configuration that results in the fewest number of test pins used. This is why the TSV-count values in Fig. 8.14 overlap even for tight test pin constraints—minimizing the TSVs used

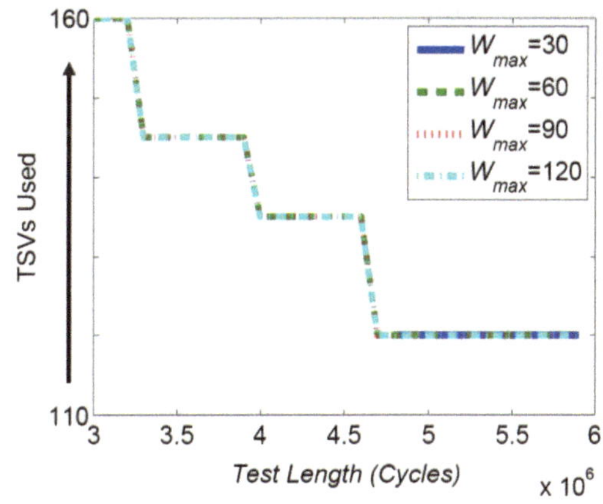

Fig. 8.15 Variation in TSVs used with T_{max} for SIC 2 with hard dies

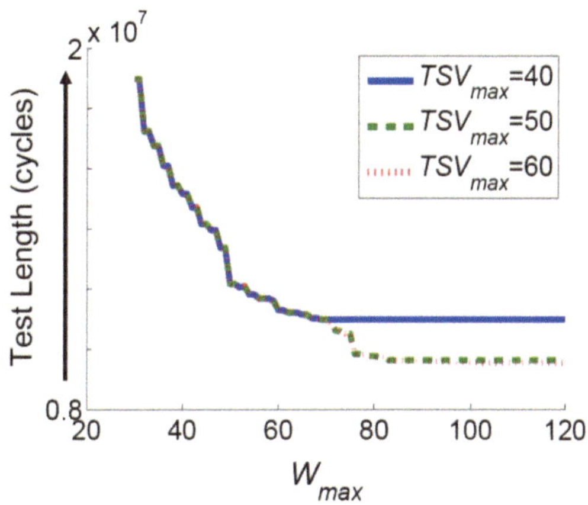

Fig. 8.16 Variation in test length with W_{max} for SIC 1 with soft dies

also tends to minimize the number of test pins used. This is seen for both SIC 1 and SIC 2 (Fig. 8.15), although the number of TSVs needed for testing in the less-optimized SIC 2 stack is higher. Results for PSHDW (3D-SIC with hard dies and test pin-use optimization) are also as expected—if minimizing TSV use also tends to minimize test pin use, then minimizing test pin use should also tend to minimize TSV use. As such, overlapping minimal test pin use for both tight and loose TSV_{max} constraints is again observed. Note that optimizing for minimum TSV or test pin use tends toward serial testing. Therefore, these optimizations result in very different test architectures than optimizing for test time. In contrast, solutions that minimize test time tend to result in the parallel testing of many dies, as can be seen for PSSD in Table 8.3.

For PSSD (3D-SIC with soft dies), Pareto-optimality is almost non-existent when W_{max} is varied; see Fig. 8.16. This is due to the fact that as dies in the stack are soft, it is always possible to find one die for which adding an extra test pin reduces the overall test length. Some Pareto-optimal points can be identified for SIC 2. This is because the most complex dies in a stack tend to be the bottleneck in reducing

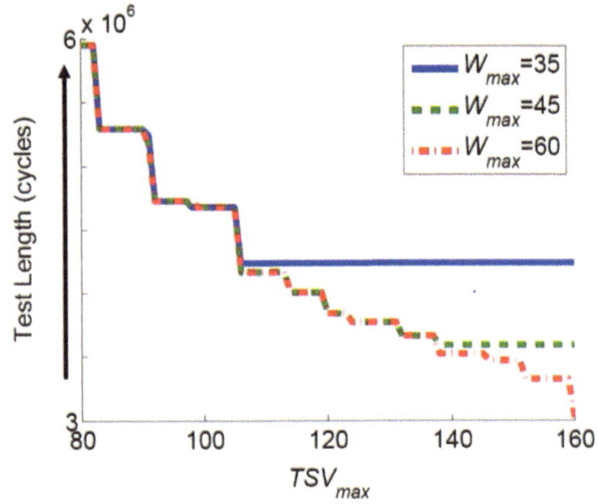

Fig. 8.17 Variation in test length with TSV_{max} for SIC 2 with soft dies

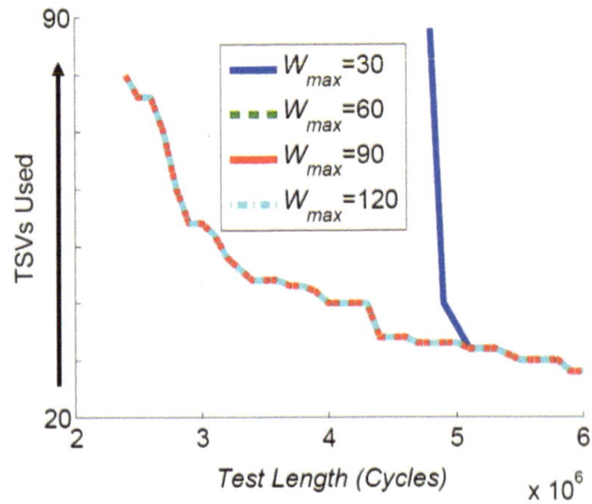

Fig. 8.18 Variation in TSVs used with T_{max} for SIC 1 with soft dies

test length. Because these dies are stacked toward the top of the stack in SIC 2, TSV constraints are more restrictive; the addition of test pins to these dies requires more TSVs and TestElevators throughout the stack. However, for PSSD, although varying W_{max} does not create Pareto-optimal points, varying TSV_{max} results in various Pareto-optimal points as shown in Fig. 8.17. Note that this effect is more pronounced in SIC 2 than in the other 3D-SICs. This is because the addition of test pins to the bottleneck die (at the highest layer) introduces a larger TSV overhead than in the other 3D-SICs. Furthermore, as long as W_{max} is sufficient, TSV_{max} is the limiter on test length. For PSHD, PSSD, and PSFD, the stack configuration (SIC 1) with the largest die at the lowest layer and the smallest die at the highest layer is the best for reducing test length while using the minimum number of TSVs. Figure 8.20 shows a comparison of optimized test lengths for soft dies between SIC 2 and SIC 3. As shown, SIC 3 leads to test lengths lower than or equivalent to SIC 2 at higher values for W_{max}. However, under tight test pin constraints SIC 2 results in better test lengths.

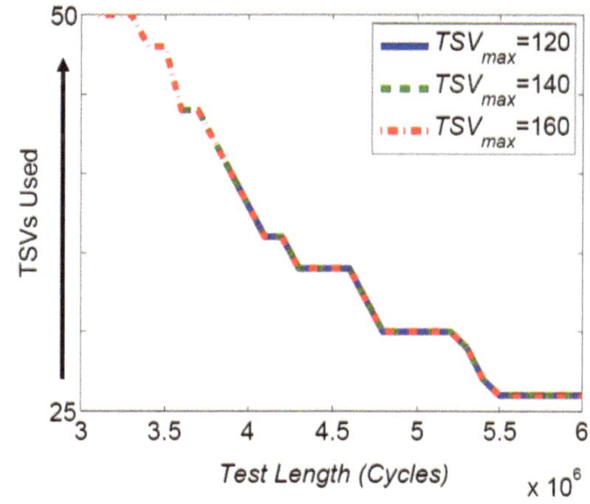

Fig. 8.19 Variation in test pins used with T_{max} for SIC 2 with soft dies

Figure 8.18 shows PSSDT (3D-SIC with soft dies and TSV-count optimization) results for SIC 1. Compared to the optimizations for hard dies, less pareto-optimality is present as expected, because more leeway exists in the soft model. For the reasons described earlier, similar results as in Figs. 8.14 are seen. Similar observations are made for test pin optimization, as seen in Fig. 8.19, which shows PSSDW (3D-SIC with soft dies and test pin-use optimization) results for and SIC 2.

Consider the optimization of SIC 2 versus SIC 3 for problem PSSD; see Fig. 8.20. For hard dies, as shown in Table 8.4, similar test times with different architectures were produced. This is because the hard die model is too limited with 3D constraints to lead to different test times for SIC 2 and SIC 3. This is not the case for soft dies, where the additional degree of freedom leads to different architectures and better test times in SIC 3 when compared to SIC 2, as expected. From a design perspective, this means that a stack layout with lower test time can be achieved if one can generally keep the most complex dies in lower layers in the stack.

Figure 8.21 demonstrates the optimization for hard dies in a 3D stack with TSV_{max} of 160 and W_{pin} of 100. As can be seen, with fixed dies TAMs there is limited opportunity for optimization given that the test lengths of each individual die cannot be altered. Thus, the test schedule has undesirable white spaces denoting wasted test resources. The use of firm and soft dies helps to eliminate wasted resources, as seen in Figs. 8.22 and 8.23, respectively. Firm dies allow for a modest reduction in test length by decreasing the number of TAM wires allotted to p22810 and utilizing those wires instead to test all the dies in parallel. With soft dies, even more test length is saved.

Finally, consider the scalability of the ILP-based optimization method by determining the stack size (number of layers) for which the ILP solver takes more than one day of CPU time. For PSHD, $M = 16$ is obtained, while for PSSD, $M = 10$. Because these values of M are unlikely for realistic stacks, it can be concluded that the proposed method is scalable and practical.

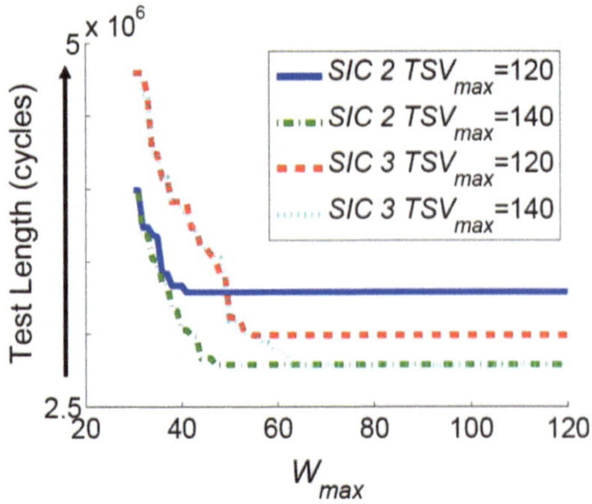

Fig. 8.20 Comparison of test lengths for PSSD for SIC 2 and SIC 3

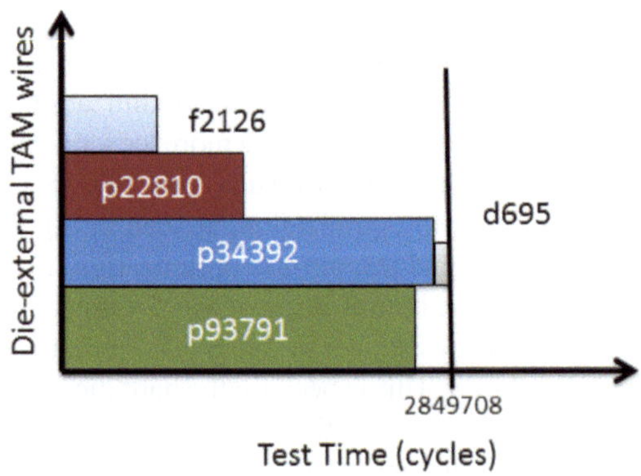

Fig. 8.21 Visualization of test schedule for SIC 1 with hard dies, $TSV_{max} = 160$, and $W_{pin} = 100$

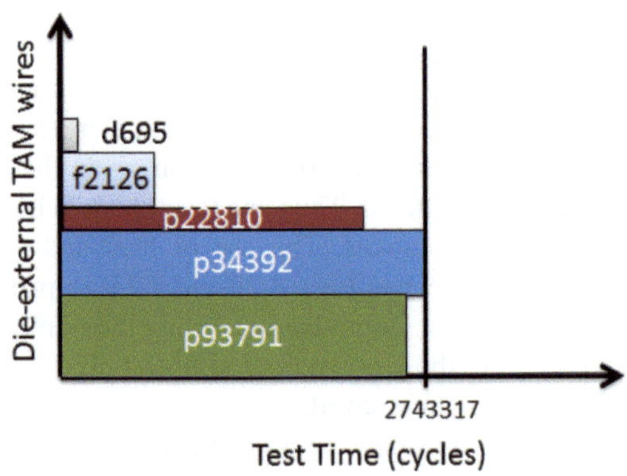

Fig. 8.22 Visualization of test schedule for SIC 1 with firm dies, $TSV_{max} = 160$, and $W_{pin} = 100$

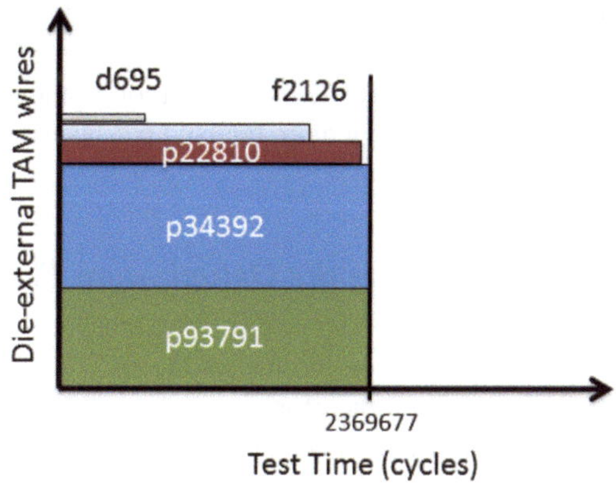

Fig. 8.23 Visualization of test schedule for SIC 1 with soft die, $TSV_{max} = 160$, and $W_{pin} = 100$

8.2.6 *Summary*

Section Summary

- An optimization based on a mathematical ILP model has been created for designing 3D and/or 2D TAMs and test schedules to minimize the test time for the final stack test.
- The optimization considered limitations on stack integration (hard, soft, and firm dies), global limits on the addition of dedicated test TSVs, and test access limitations to dies in the stack.
- The ILP implementation generally resulted in greater reduction in test time compared to a greedy algorithm. This is particularly true for the more computationally complex problem of co-optimizing 2D and 3D TAMs and test schedules.
- Pareto-optimality is shown for integration of hard dies, while increasing test resources tend to result in reductions in test time for soft-die integration.
- Increasing the availability of external test pins tends to have a greater impact on test time than easing TSV limitations, although both are important factors in reducing test time.

8.3 Extending Test Optimization for Multiple Test Insertions and Interconnect Test

In Sect. 8.2, optimization techniques were introduced for minimizing the test time for final stack test. A global limit was set on the number of dedicated TSVs to be used for test access and constraints were imposed on test bandwidth due to a limited number of test pins on the lowest die in the stack. This section extends the mathematical model to allow for the optimizer to consider any or all post-bond test insertions desired by the 3D test engineer. This is a more general approach that still allows for optimization for the final stack test alone, or any number of post-bond test insertions. Furthermore, the test times for TSVs and die-external logic will be considered in the extended optimization framework. Finally, TSV limits for the 3D TAM are applied on a more realistic per-layer basis.

8.3.1 Modifying the Optimization Problem Definition

As mentioned earlier, the lowest die in a 3D SIC is usually directly connected to chip I/O pins, and therefore it can be tested using package pins. To test the other dies in the stack, TAMs that enter the stack from the lowest die should be provided. To transport test data up and down the stack, *test elevators* must be included on each die except for the highest die in the stack [42]. The number of test pins and test elevators, as well as the number of TSVs used, affect the total test time for the stack.

Many new manufacturing steps are needed for the production of 3D-SICs than for 2D-SICs, including TSV creation, wafer thinning, alignment, and bonding. These steps can introduce possible defects that do not arise for 2D-SICs [43]. Such defects include incomplete fill of TSVs, misalignment, peeling and delamination, and cracking due to back-side grinding. It is difficult to carry out pre-bond testing of TSVs due to limitations in probe technology and the need for contactless probing. Thus, post-bond partial-stack and in-stack TSV testing are needed to detect these new defects and reduce defect escapes.

The following two examples, for hard dies and soft dies, respectively, highlight the limitations of optimization techniques that are oblivious to multiple test insertions. In Sect. 8.2, optimization decisions were made considering only the final stack test after all dies have been bonded. These models cannot be directly applied to optimize for multiple test insertions, as shown in Fig. 8.24.

Example 1: Consider an attempt to optimize for two test insertions, the first with three dies on the stack as in Fig. 8.24a and the second with all four dies as in Fig. 8.24b, by building upon the test architecture created for the first test insertion. A global TSV limit of 90 is imposed for this example. As seen in Fig. 8.24a, Die 1 is tested in series with Die 2 and Die 3, resulting in a minimized test time of 1500 cycles and the use of 80 TSVs as test elevators. It is then attempted to add the fourth die while preserving the previous architecture, but this leads to a violation of the TSV limit because 100 TSVs are now needed (the test elevators marked in the figure exceed the mandated upper limit). If instead optimization starts from the complete stack and works backwards, suboptimal results for the test time are obtained. This is because the architecture created for the final stack test cannot support optimal test schedules for other intermediate test insertions. Therefore, new optimization techniques are needed for partial stack testing. The following example highlights this problem.

Example 2: It can also be shown that optimizing only for the final stack test does not result in optimum test times for multiple test insertions. Consider an SIC with

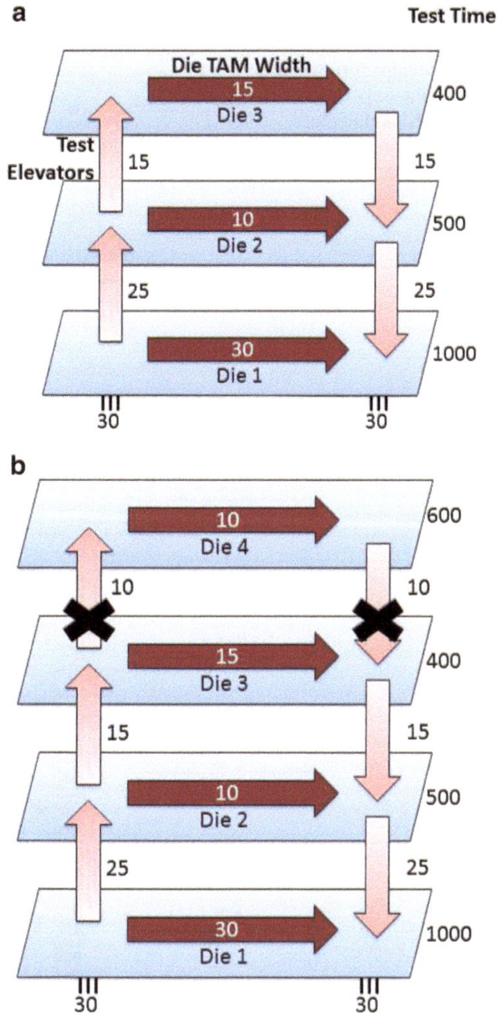

Fig. 8.24 Example 1: 3D-SIC with four hard dies: (**a**) test architecture for test insertion of the 3-die partial stack without considering multiple test insertions and (**b**) TSV constraint violation when considering complete stack test

three dies from the ITC'02 SOC Benchmarks [19] as shown in Fig. 8.25. There are 40 test pins available to the stack and a limit of 40 TSVs per side of each die as the maximum number of TSVs to use for the test infrastructure. The test time for each die is determined by its test architecture, which in this example relies on daisychaining.

Figure 8.25a shows the resulting test elevator widths and the number of TSVs used if the stack is optimized to reduce the test time of the final stack test after all dies have been bonded. This architecture allows for the testing of all three dies in parallel, which results in a final stack test time of 1313618 cycles with individual die test times of 1313618, 1303387, and 658834 cycles, from top to bottom. The architecture uses 38 out of the 40 allowed test pins and 64 TSVs. When both possible

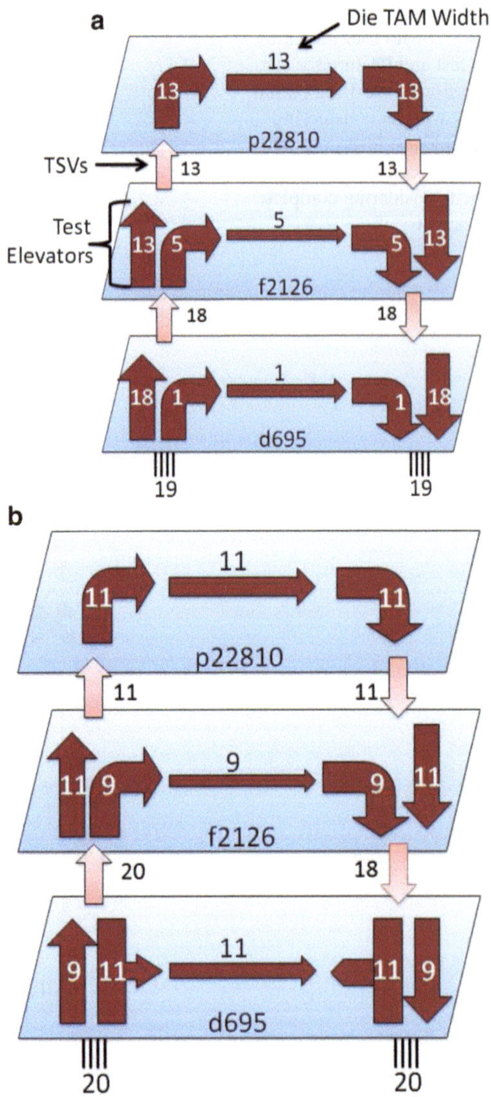

Fig. 8.25 Example 2: 3D-SIC with three soft dies: (**a**) test architecture when considering only final stack test and (**b**) architecture optimized for all post-bond test insertions

stack tests are considered, with the first stack test done after the middle die is bonded to the lower die and the final stack test, the total test time becomes 2617005 cycles (Die 1 and Die 2 are tested in parallel for the first stack test).

Figure 8.25b shows the test architecture created if all stack tests are considered during optimization, which uses all 40 test pins and 62 TSVs. This architecture allows Die 1 and Die 2 to be tested in parallel for the first stack test, and then Die 2 and Die 3 to be tested in parallel after Die 1 is tested for the final stack test. The test times for the dies from top to bottom are 1338480, 700665, and 75004, respectively. This results in a final stack test time of 1413484 cycles, an increase in time over the previous example. However, when considering all stack tests, this architecture results in a total test time of 2114149 cycles (700665 cycles for the first stack test), a considerable reduction over the previous example. This example clearly shows the impact of the optimization target on the architecture and test time. Therefore, a test-architecture optimization algorithm for 3D-SICs has to minimize the test time while taking into account 3D design constraints as well as all of the post-bond tests that will be run on the 3D-SIC. The resulting optimized 3D TAM must also allow for different test schedules for each post-bond test.

With the above examples as motivation for better optimization methods, the problems addressed in this section can be formally stated. The problem of test-architecture optimization for 3D-SICs with hard dies for all stack test insertions is defined as follows.

3D-SIC with Hard Dies and Multiple Test Insertions (P^H_{MTS})
Givens include a stack with a set M of dies and total number of test pins W_{max} available for test. For each die $m \in M$, its tier number l_m in the stack, the number of test pins w_m ($w_m \leq W_{max}$) required to test the die, the associated test time t_m, and a maximum number of TSVs ($TSVmax_m$) that can be used for TAM design between die $m-1$ and m ($m > 1$) are given. The goal is to determine an optimal TAM design for the stack and test schedule for each stage of stacking such that the total test time T, i.e., the sum of the test times for all the desired stack tests (final stack test or multiple test insertions), is minimized and the number of TSVs used per die does not exceed $TSVmax_m$.

For a 3D-SIC with soft dies, the test-architecture for each die is not pre-defined, but is determined during the test-architecture design for the stack. This scenario provides greater flexibility in terms of test-time optimization. The problem of test-architecture optimization for 3D-SICs with soft dies and multiple test insertions is formally defined as follows.

3D-SIC with Soft Dies (P^S_{MTS})
Givens include a stack with a set M of dies and the total number of test pins W_{max} available for test. For each die $m \in M$, its tier number l_m in the stack, a maximum number of TSVs ($TSVmax_m$) that can be used for TAM design between die $m-1$ and m ($m > 1$), and the total number of cores c_m are given. Furthermore, for each core n, the number of inputs i_n, outputs o_n, total number of test patterns p_n, total number of scan chains s_n, and for each scan chain k, the length of the scan chain in flip flops $l_{n,k}$ are given. The goal is to determine an optimal TAM design and test schedule for each stage of stacking, as well as for each die, such that the total test time T for the stack is minimized and the number of TSVs used per die does not exceed $TSVmax_m$.

Consider further the testing of TSVs and die-external logic along with the cores in each die. There are two variants to this problem; see Fig. 8.26. Figure 8.26a shows a three-die stack with a test architecture using fat wrappers to allow TSV tests to take place in parallel with module testing on the dies. Each die has a separate TAM for die-external tests, both to higher and lower dies in the stack. Each TAM has its own width and utilizes different test pins. In this case, test pins used for TSV testing are not available for die testing. For example, internal test of Die 2 and external test of

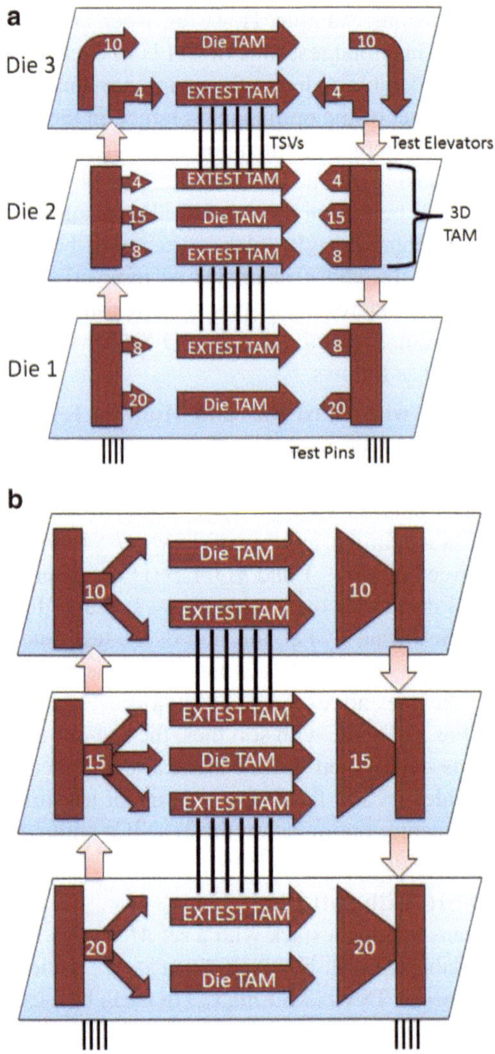

Fig. 8.26 Example of a test architecture for a 3D-SIC including die-external tests with (**a**) fat wrappers and (**b**) thin wrappers

the TSVs and die-external logic between Die 2 and Die 3 can occur simultaneously, utilizing a total of 23 test pins. The second variant uses a thin wrapper which allows die-external tests to only take place serially with module testing on the die as seen in Fig. 8.26b. This design allows the TAMs for die-external test to access the full test width for the die, but test pins are shared between all TAMs on the die. In this architecture, the external tests for logic and TSVs between Die 2 and Die 3 can be carried out in parallel with the internal test of Die 1, but not in parallel with any other test.

Below, the problem definitions for these two variants is presented, where "||" refers to parallel and "−−" refers to serial. The problems of testing hard dies including TSV tests for a single test insertion (final stack test only) are stated below as Problem 3(a) and Problem 3(b), respectively.

3D-SIC with Hard Dies and (a) fat wrapper ($P^{H}_{DTSV,||}$) or (b) thin wrapper ($P^{H}_{DTSV,--}$)
A stack with a set M of dies and the total number of test pins W_{max} available for test is given. For each die $m \in M$, its tier number l_m in the stack, the number of test pins w_m ($w_m \leq W_{max}$) required to test the die, the associated test time t_m, and a maximum number of test elevators ($TSVmax_m$) that can be used for TAM design between die $m-1$ and m are given. Furthermore, the number of functional TSVs (Tf_m) and test width given to the TSVs (Wt_{max}) between die $m-1$ and m, and the number of test patterns (Pf_m) for the functional TSVs between die $m-1$ and m (for $m > 1$) are given. The goal is to determine an optimal TAM design and a test schedule for the stack for die-internal and die-external tests such that the total test time T is minimized and the number of test TSVs used per die does not exceed $TSVmax_m$.

The problems of testing soft dies including TSV tests for a single test insertion are stated below as Problem 4(a) and Problem 4(b).

3D-SIC with Soft Dies and (a) fat wrapper ($P^{S}_{DTSV,||}$) or (b) thin wrapper ($P^{S}_{DTSV,--}$)
A stack with a set M of dies and the total number of test pins W_{max} available for test is given. For each die $m \in M$, its tier number l_m in the stack, a maximum number of test elevators ($TSVmax_m$) that can be used for TAM design between die $m-1$ and m, and the number of functional TSVs (Tf_m) between die $m-1$ and die m are given. Furthermore, the number of test patterns (Pf_m) for the functional TSVs between die $m-1$ and m (for $m > 1$), and the total number of cores c_m are given. For each core n, the number of inputs i_n, outputs o_n, total number of test patterns p_n, total number of scan chains s_n, and for each scan chain k, the length of the scan chain in flip flops $l_{n,k}$ are given. Determine an optimal TAM design for the stack and TSVs and a test schedule for the stack such that the total test time T is minimized and the number of test TSVs used per die does not exceed $TSVmax_m$.

All six problems presented above are NP-hard from "proof by restriction" [21], as they can be reduced using standard techniques to the rectangle packing problem, which is known to be NP-hard [16]. For example, for Problem P^{S}_{MTS}, if the constraints related to maximum number of TSVs are removed and only the final stack test insertion is considered, each die can be represented as a set of rectangles with different widths and heights, where width is equal to its test time for a given TAM width and height equal to the number of required test pins. Now, all these rectangles (dies) must be packed into a bin with width equal to the total number of test pins and height equal to the total test time for the stack, which must be minimized. Despite the NP-hard nature of these problems, they can be solved optimally because the number of layers in a 3D-SIC is expected to be limited, e.g., up to four layers have been predicted for logic stacks [17].

8.4 Derivation of the Extended ILP Model

In this section, ILP is used to model and solve the problems defined in the previous section. The problem instances in practice are relatively small for realistic stacks with anywhere from two to eight dies. Therefore, ILP methods are good candidates for solving these optimization problems.

8.4.1 ILP Formulation for Problem P^{H}_{MTS}

To create an ILP model for this problem, the set of variables and constraints must be defined. To begin, define a binary variable x_{ijk}, which is equal to 1 if die i is tested in parallel with die j for a test insertion when there are k die in the stack, and 0 otherwise. There are $M - 1$ test insertions, one for each additional die added to the stack, such that k ranges from 2 to M. Constraints on variable x_{ijk} can be defined as follows:

$$x_{iik} = 1 \qquad \forall k, i \leq k \tag{8.31}$$

$$x_{ijk} = x_{jik} \qquad \forall k, \{i, j\} \leq k \tag{8.32}$$

$$1 - x_{iqk} \geq x_{iqk} - x_{jqk} \geq x_{ijk} - 1 \qquad \forall k, \{i, j, q\} \leq k, i \neq j \neq q \tag{8.33}$$

The first constraint indicates that every die is always considered to be tested with itself for every test insertion. The second constraint states that if die i is tested in parallel with die j for insertion k, then die j is also tested in parallel with die i for insertion k. The last constraint ensures that if die i is tested in parallel with die j for insertion k, then it must also be tested in parallel with all other dies that are tested in parallel with die j for insertion k.

Next, define a second binary variable y_{ik}, which is equal to 0 if die i is tested in parallel with die j on a lower layer ($l_i > l_j$) for insertion k, and 1 otherwise. The total test time T for the stack is the sum of the test times of all dies that are tested in series plus the maximum of the test times for each of the sets of parallel tested dies for all test schedules at every test insertion. Using variables x_{ijk} and $y_i k$, the total test time T for all test insertions with the set of dies M can be defined as follows.

$$T = \sum_{k=2}^{|M|} \sum_{i=1}^{k} y_{ik} \cdot \max_{j=i..k} \{x_{ijk} \cdot t_j\} \tag{8.34}$$

It should be noted that Equation (8.34) has two non-linear elements, the *max* function, and the product of $y_i k$ variable and the *max* function. This is linearized by introducing two new variables. The variable c_{ik} takes the value of the max function for each die i for test insertion k and the variable u_{ik} represents the product $y_{ik} \cdot c_{ik}$. The variables u_{ik} and c_{ik} are defined using standard linearization techniques. The linearized function for total test time can be written as follows.

$$T = \sum_{k=2}^{|M|} \sum_{i=1}^{k} u_{ik} \tag{8.35}$$

As the number of test pins used for parallel testing of dies should not exceed the given test pins W_{max} across all test schedules for every test insertion, a constraint on the total number of pins used to test all dies in a parallel set in any given test insertion can be defined as follows for all k.

$$\sum_{j=1}^{k} x_{ijk} \cdot w_j \leq W_{max} \qquad \forall i \leq k \tag{8.36}$$

Similarly, the total number of used TSVs should not exceed the given TSV limit ($TSVmax_m$) for each die face across all test insertions. It should be noted that $TSVmax_2$ is the limit for the upper face of die 1 and the lower face of die 2, $TSVmax_3$ is for the upper face of die 2 and lower face of die 3, and so forth. The number of TSVs used to connect layer i to layer $i-1$ is the maximum of the number of pins required by the layer at or above layer i that takes the most test pin connections, and the sum of parallel-tested dies at or above layer i in the same parallel tested set across all test insertions. Based on this, the constraint on the total number of TSVs used in a test architecture can be defined as follows.

$$\max_{i<=k<=|M|} \{w_i, \sum_{j=i}^{k} w_j \cdot x_{kj}\} \leq TSVmax_i \qquad \forall i \geq 2, \tag{8.37}$$

The above set of constraints can be linearized by representing the max function with a variable d_i. Finally, to complete the ILP model for Problem P_{MTS}^{H}, constraints on the binary variable y_{ik} and the relationship between binary variables y_{ik} and x_{ijk} must be defined. For this purpose, a constant C is introduced that approaches but is less than 1. The variable y_{ik} can then be defined as follows:

$$y_{1k} = 1 \qquad \forall k \tag{8.38}$$

$$y_{ik} \geq \frac{1}{1-i} \sum_{j=1}^{i-1} (x_{ijk} - 1) - C \qquad \forall k, i \leq k, i \neq 1 \tag{8.39}$$

The first equation forces y_{1k} to 1, because the lowest layer cannot be tested in parallel with any layer lower than itself. Constraint 8.39 defines y_{ik} for the other layers. To understand this constraint, first make the observation that the objective function (as shown in Equation (8.34)) would be minimized if each y_{ik} is zero. This would make the objective function value equal to 0, which is an absolute minimum test time. Thus, y_{ik} must be restricted to 1 only where it is absolutely necessary, and then the objective function can be relied upon to assign a value 0 to all unrestricted

Fig. 8.27 ILP model for the 3D TAM optimization Problem P^{H}_{MTS}

Objective:

$$\text{Minimize } \sum_{k=2}^{|M|} \sum_{i=1}^{k} u_{ik}$$

Subject to the Constraints:

$$t_{max} = \max_{i=1}^{|M|} t_i$$
$$c_{ik} \geq x_{ijk} \cdot t_j \qquad \forall k, \{i,j\} \leq k$$
$$u_{ik} \geq 0 \qquad \forall k, i \leq k$$
$$u_{ik} - t_{max} \cdot y_{ik} \leq 0 \qquad \forall k, i \leq k$$
$$u_{ik} - c_{ik} \leq 0 \qquad \forall k, i \leq k$$
$$c_{ik} - u_{ik} + t_{max} \cdot y_{ik} \leq t_{max} \qquad \forall k, i \leq k$$
$$\sum_{j=1}^{k} x_{ijk} \cdot w_j \leq W_{max} \qquad \forall k, i \leq k$$
$$x_{iik} = 1 \qquad \forall k, i \leq k$$
$$x_{ijk} = x_{jik} \qquad \forall k, \{i,j\} \leq k$$
$$1 - x_{iqk} \geq x_{iqk} - x_{jqk} \geq x_{ijk} - 1 \quad \forall k, \{i,j,q\} \leq k, i \neq j \neq q$$
$$d_i \leq TSVmax_i \qquad \forall i \neq 1$$
$$d_i \geq \sum_{j=i}^{k} w_j \cdot x_{ijk} \qquad \forall i \neq 1, k \geq i$$
$$d_i \geq w_j \qquad \forall i, j = i..|M|$$
$$y_{1k} = 1 \qquad \forall k$$
$$y_{ik} \geq \frac{1}{1-i} \sum_{j=1}^{i-1} (x_{ijk} - 1) - C \qquad \forall k, i \leq k, i \neq 1$$

y_{ik} variables. This equation considers the range of values that the sum of x_{ijk} can take. The fraction in the equation normalizes the sum to a value between 0 and 1 inclusive, while the summation considers all possible cases for a die being in parallel with dies below it. The complete ILP model is shown in Fig. 8.27.

Figure 8.28 illustrates the ILP model for P^{H}_{MTS}. The 3D-SIC in Fig. 8.28 consists of three dies, with test times of 1333098, 700665, and 106391 cycles moving up the stack. There are 22 test pins available, so W_{max} is 22. TSV_{max} is set to 20, such that there can be no more than 20 dedicated test TSVs between any two dies (this limits the TSVs per die to 40). There are two test insertions, the first when Die 1 and Die 2 are stacked and the second for the complete stack. In the first test insertion (k = 2), the optimal solution that Die 1 and Die 2 are tested in parallel is calculated. As such, $x_{1,1,2}$, $x_{1,2,2}$, $x_{2,1,2}$, and $x_{2,2,2}$ are all equal to 1. Die 2 is tested in parallel with a die below it (Die 1), so $y_{2,2}$ is 0 and $y_{1,2}$ is 1. The test time for this test insertion is 1333098 cycles, because $u_{1,2}$ is 1333098 and $u_{2,2}$ is 0.

For the second test insertion (k = 3), the optimal solution is to test Die 1 and Die 2 in parallel, and then to test Die 3. Because Die 1 and Die 2 are tested in parallel again, $x_{1,1,3}$, $x_{1,2,3}$, $x_{2,1,3}$, and $x_{2,2,3}$ are equal to 1. For Die 3, $x_{3,3,3}$ is 1. All other x_{ijk} variables are 0. As before, $y_{2,3}$ is 0 since Die 2 is tested in parallel with Die 1. The variables $y_{1,3}$ and $y_{3,3}$ are 1. The test time for this test insertion is 1439489, as $u_{1,3}$ is 1333098, $u_{2,3}$ is 0, and $u_{3,3}$ is 106391. In this architecture, d_2 is 20 and d_3 is 14, neither of which violate the TSV limit. All 44 test pins are utilized.

The above ILP model is a generalization of the special case presented in Sect. 8.2, in which test-time was minimized for only the final stack test. If the variable k is constrained to take only one value, namely M, in P^{H}_{MTS}, then optimization will produce a test architecture and test schedule that minimizes test time only for the final stack, i.e., the objective in Sect. 8.2. Optimization for only the final stack test

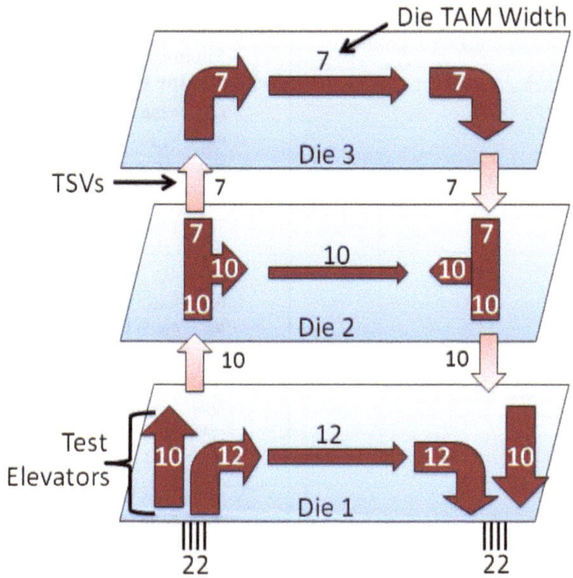

Fig. 8.28 Example of a test architecture in an optimized 3D-SIC with hard dies

is referred to as P_{FS}^{H} and P_{FS}^{S} for hard and soft dies, respectively. Therefore, an advantage of the optimization model proposed here is that it is flexible—it can be easily tailored to minimize test time for any number of stack tests, from one final test to two or more intermediate test insertions. Multiple options can be automatically generated for testing the stack. For example, suppose there is interest in two test insertions—after the second die is bonded to the first die and the final stack test. By allowing k to now take two values, namely 2 and M, the test time for the stack can be minimized by considering only these two insertions.

8.4.2 ILP Formulation for Problem P_{MTS}^{S}

The ILP formulation for 3D-SICs with soft cores and multiple test insertions is derived in a similar manner as the 3D-SIC with hard cores above. In this case, the test time t_i for die i is a function of the TAM width w_i assigned to it. Using the variables x_{ijk} and $y_i k$ as defined in Sect. 8.4.3, the total test time T for the stack with the set of soft dies M across all test insertions can be defined as follows.

$$T = \sum_{k=2}^{|M|} \sum_{i=1}^{k} y_{ik} \cdot \max_{j=i..k} \{x_{ijk} \cdot t_j(w_j)\} \tag{8.40}$$

It should be noted that Equation (8.40) has several non-linear elements. To linearize this equation, first define the test time function. For this purpose, a binary variable

Fig. 8.29 ILP model for 3D TAM optimization Problem P^S_{MTS}

Objective:

Minimize $\sum^(|M|)_{k=2}\sum^{k}_{i=1} u_i$

Subject to the Constraints:

$$
\begin{array}{ll}
t_{max} = \max_{i=1}^{|M|} t_i & \\
c_{ik} \geq v_{ijk} & \forall k, i \leq k, i \leq j \leq k \\
v_{ijk} \geq 0 & \forall k, \{i,j\} \leq k \\
v_{ijk} - t_{max} \cdot x_{ijk} \leq 0 & \forall k, \{i,j\} \leq k \\
-\sum_{n=1}^{k_i} (g_{jn} \cdot t_j(n)) + v_{ijk} \leq 0 & \forall k, \{i,j\} \leq k \\
\sum_{n=1}^{k_i} (g_{jn} \cdot t_j(n)) - v_{ijk} + t_{max} \cdot x_{ijk} \leq t_{max} & \forall k, \{i,j\} \leq k \\
u_{ik} \geq 0 & \forall k, i \leq k \\
u_{ik} - t_{max} \cdot y_{ik} \leq 0 & \forall k, i \leq k \\
u_{ik} - c_{ik} \leq 0 & \forall k, i \leq k \\
c_{ik} - u_{ik} + t_{max} \cdot y_{ik} \leq t_{max} & \forall k, i \leq k \\
z_{ijkj} \geq 0 \forall k, \{i,j\} \leq k & \\
z_{ijkj} - t_{max} \cdot x_{ijk} \leq 0 \forall k, \{i,j\} \leq k & \\
-wj + z_{ijkj} \leq 0 \forall k, \{i,j\} \leq k \forall k, \{i,j\} \leq k & \\
w_j - z_{ijkj} + t_{max} \cdot x_{ijk} \leq t_{max} \forall k, \{i,j\} \leq k & \\
\sum_{i=2}^{|M|} d_i \leq TSVmax_i & \\
d_q \geq \sum_{j=i}^{k} z_{ijkj} & \forall 2 \leq q \leq |M|, k \geq q, q \leq i \leq k \\
d_i \geq w_j & \forall i, j \geq i \\
\sum_{j=1}^{k} z_{ijkj} \leq W_{max} & \forall k, i \leq k \\
x_{iik} = 1 & \forall k, i \leq k \\
x_{ijk} = x_{jik} & \forall k, \{i,j\} \leq k \\
1 - x_{iqk} \geq x_{iqk} - x_{jqk} \geq x_{ijk} - 1 & \forall k, \{i,j,q\} \leq k, i \neq j \neq q \\
y_{1k} = 1 & \forall k \\
y_{ik} \geq \frac{1}{1-i} \sum_{j=1}^{i-1} (x_{ijk} - 1) - C & \forall k, i \leq k, i \neq 1
\end{array}
$$

g_{in} is introduced where $g_{in} = 1$ if $w_i = n$, and 0 otherwise. This expression is then linearized using the variable v_{ijk} for $x_{ijk} \cdot \sum_{n=1}^{k_i} (g_{jn} \cdot t_j(n))$. It should be noted that, although the variable x_{ijk} can change for different test insertions depending on the test schedule for each insertion, there must be a single value of test pins used for each die to reflect an architecture that can support all test schedules used throughout all test insertions. Similar to Equation (8.35), the variable c_{ik} takes the value of the max function for each die i for each test insertion k and the variable u_{ik} represents the product $y_{ik} \cdot c_{ik}$. Because w_j is now a decision variable, $x_{ijk} \cdot w_j$ is linearized using a new variable z_{ijkj}. The max function is represented by the variable d_i as before. By using the variable z_{ijkj}, the TAM width that can be given to each die can be constrained by an upper limit, which is the number of available test pins. This is represented by the following inequality.

$$\sum_{j=i}^{k} z_{ijkj} \leq W_{max} \qquad \forall k, i \leq k \tag{8.41}$$

The complete ILP model for Problem P^S_{MTS} is shown in Fig. 8.29.

8.4.3 ILP Formulations for $P^H_{DTSV,||}$, $P^H_{DTSV,--}$, $P^S_{DTSV,||}$, and $P^S_{DTSV,--}$

The ILP formulations in Sects. 8.4.1 and 8.4.2 do not account for the TAM architecture and test time needed to test the TSVs. In order to include TSV tests, the TSV connections between two layers are treated as a "virtual" die. This die has a predefined test width for hard dies or a range of possible widths with associated test time for soft dies, just as an actual die for each model. In order to calculate the test time for the TSV "layer", the formula $\lceil Tf_m/\lfloor(pins/4)\rfloor\rceil(Pf_m + 1)$ is used, where $pins$ refers to the number of test pins available for TSV testing.

The above expression begins by taking the floor of $pins/4$. It is assumed that, between two dies, there is any number of functional TSVs that are used for either input to the dies or output from the dies. The TSV tips are latched and the corresponding flip-flops are connected to form one or more scan-chains. Thus, the scan-flops are also assumed to be unidirectional. In order to quickly test all TSVs, shift-in, shift-out, and capture on both sides of the TSVs must be allowed for simultaneously. This requires 4 test pins for each TSV scan-chain, i.e., an input and output pin for each scan-chain for both sides of the TSVs as seen in Fig. 8.30a. As can be seen from Fig. 8.30b, there is no reduction in test time for stack tests if unequal numbers of TSV scan-chains for dies on either side of the TSVs are considered, as the bottleneck in test time would then be the die with the fewest TSV scan-chains (TSV testing requires use of TSV scan-chains on both sides of the TSVs in parallel). At least four more test pins must be added for a reduction in test time as shown in Fig. 8.30c. Thus, test pin use is evenly divided among the dies for TSV tests, and the number of scan-chains on either side of the TSVs is $\lceil Tf_m/\lfloor(pins/4)\rfloor\rceil$. This is multiplied by the number of patterns required for TSV testing plus one to accommodate for shift-in and shift-out operations. Without loss of generality, the number of test pins required for control signals to a die-level wrapper for TSV testing is not considered.

Figure 8.31 shows the difference between fat and thin wrappers. In Fig. 8.31a, the die-internal and die-external TAMS utilize different test pins. Thus, EXTEST can be performed in parallel with either or both INTEST of Die 1 and Die 2. This is representative of fat wrappers. For thin wrappers, Fig. 8.31b shows that the same test pins are utilized for die-internal and die-external tests, and as such test data must be multiplexed to the correct TAM. In this case, the INTEST of each die can be performed in parallel, but EXTEST cannot be performed in parallel with INTEST on Die 1 or Die 2 (although it can be performed in parallel with INTEST of other dies in the stack).

The ILP formulation for 3D-SICs with hard cores including TSVs is derived in a similar manner as the 3D-SIC with hard cores in Problem 1. Begin by removing the subscript k from all variables that account for multiple test insertions—only consider the final stack test. As stated before, consider a set of TSVs between two layers to be a virtual die for purposes of optimization, such that Die 1 is the lowest

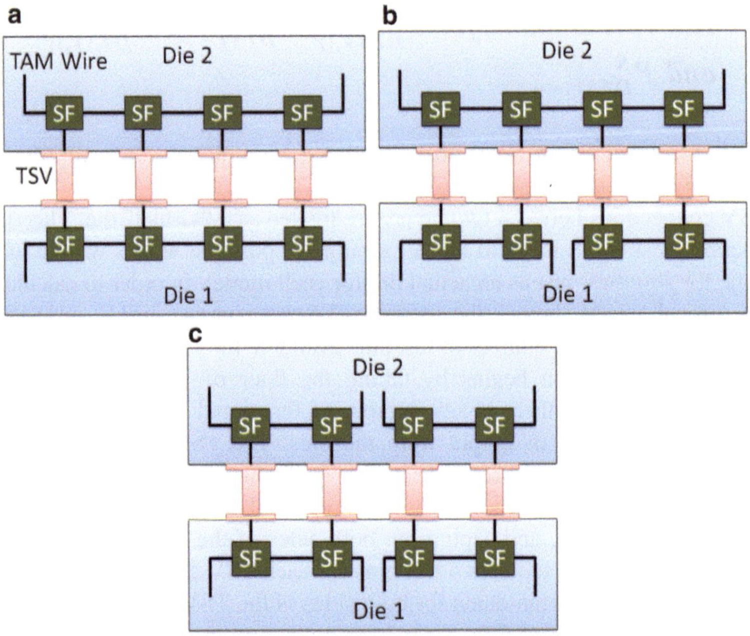

Fig. 8.30 Example of scan chains for die-external tests: (**a**) one scan chain per die with a four scan-flop bottleneck, (**b**) one scan chain on one die and two scan chains on the other with a four scan-flop bottleneck, and (**c**) two scan chains per die with a two scan-flop bottleneck

die in the stack, Die 2 represents the TSVs between Die 1 and Die 3, Die 3 is the second die in the stack, Die 4 represents the TSVs between Die 3 and Die 5, and so on. In this way, odd-numbered dies are actual dies and even-numbered dies represent the TSVs between two odd-numbered dies.

Variables and constraints must be added in order to accurately model the dies representing TSVs. Begin by defining the variable $p_i = 1$ if a die i representing TSVs can be tested in parallel with the dies below and above it and 0 otherwise. In the case of fat wrappers, p_i is left as a decision variable and in the case of thin wrappers, p_i is forced to 0 for all dies representing TSVs. For actual (even-numbered) dies, p_i is always 1. If a die representing TSVs can be tested in parallel with the dies around it, then the number of test pins given to TSV testing reduces the number of pins available to the dies for testing. Otherwise, the TSVs can use all the test pins that are utilized by the dies around it. Define the set die to contain all even-numbered dies, or the actual dies, in the stack. The following restrictions on p_i are added for fat wrappers (the first constraint applies for thin wrappers as well):

$$p_i = 1 \qquad \forall i \in die \tag{8.42}$$

$$p_{i+1} \leq w_i - w_{i+1} \qquad \forall i \in die, i \neq |M| \tag{8.43}$$

$$p_{i+1} \leq w_{i+2} - w_{i+1} \qquad \forall i \in die, i \neq |M| \tag{8.44}$$

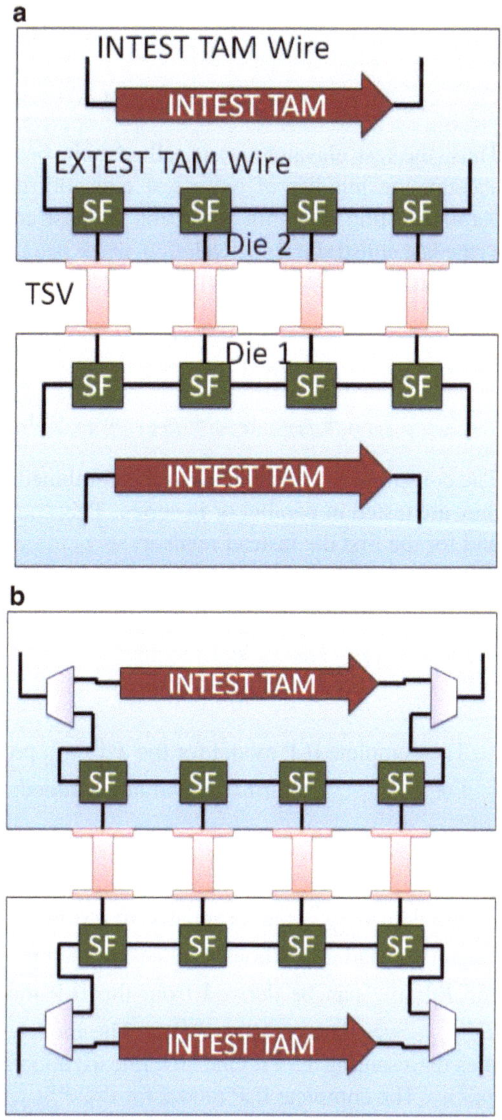

Fig. 8.31 Simplified illustration of (**a**) fat wrappers and (**b**) thin wrappers

This restricts p_i to 0 if all pins on a die are shared for TSV tests, and otherwise leaves the variable open to optimization. The variable p_i is used to accurately represent the number of TSVs and test pins utilized in the stack and uses these constraints to further restrict the variable p_i. The variable d_i is redefined as follows:

$$d_i \geq \sum_{j=k}^{|M|} w_j \cdot x_{kj} \qquad \forall i \in die, i \neq |M|, k = i..|M| \tag{8.45}$$

$$d_i \geq w_j \qquad \forall i \in die, i \neq |M|, j = i..|M| \tag{8.46}$$

$$d_i \geq w_j + p_{j-1} \cdot w_{j-1} + p_{j+1} \cdot w_{j+1} \quad \forall i \in die, j \geq i \tag{8.47}$$

Thus, the test elevator use is tallied only between actual dies, while taking into account the number of extra test elevators used for the TSV layer considering parallel testing. For thin wrappers, the last constraint on d_i is removed. For Die 1, the last constraint is instead $d_i \geq w_j + p_{j+1} \cdot w_{j+1}$. Test pin constraints therefore reduce to:

$$\sum_{j=1}^{|M|} x_{ij} \cdot w_j \leq W_{max} \qquad \forall i \tag{8.48}$$

$$w_i + p_{i-1} \cdot w_{i-1} + p_{i+1} \cdot w_{i+1} \leq W_{max} \qquad \forall i \in die, i \neq |M| \tag{8.49}$$

The constraint 8.51 accounts for the combined test pin use by dies and TSVs when they are tested in parallel or in series. This constraint is removed for thin wrappers, and for the first die instead reads $w_i + p_{i+1} \cdot w_{i+1} \leq W_{max}$, $\forall i \in die, i \neq |M|$. It is necessary to further update the variable x_{ij} as follows:

$$x_{i,i+1} \leq p_{i+1} \qquad \forall i \in die, i \neq |M| \tag{8.50}$$

$$x_{i+2,i+1} \leq p_{i+1} \qquad \forall i \in die, i \neq |M| \tag{8.51}$$

The complete ILP model for the $P^H_{DTSV,||}$ problem is given in Fig. 8.32.

For $P^H_{DTSV,--}$, a further constraint is added:

$$x_{i-1,i+1} = 0 \quad \forall i \in die, i \neq \{1, |M|\} \tag{8.52}$$

Problem $P^S_{DTSV,||}$ is formulated in a similar manner as P^S_{MTS}, with the subscript k removed and added constraints as defined in $P^H_{DTSV,||}$. In this way, the ILP model for $P^S_{DTSV,||}$ can be derived from the ILP model for P^S_{MTS} in the same way as $P^H_{DTSV,||}$ was derived from P^H_{MTS}. The model for $P^S_{DTSV,--}$ forces all p_i to 0 for dies representing TSVs. One new linearization constraint, pw_i is added to represent $p_i \cdot w_i$. The complete ILP model for the $P^S_{DTSV,||}$ problem is given in Fig. 8.33.

8.5 Results and Discussion for the Multiple-Test Insertion ILP Model

In this section, experimental results are presented for the ILP models given in Sect. 8.4. SIC 1 and SIC 2 as shown in Fig. 8.9 from Sect. 8.2 are utilized as benchmarks for this section. In SIC 1, the most complex die (p93791) is placed at the bottom, with die complexity decreasing as one moves up the stack. The order is reversed in SIC 2.

Fig. 8.32 ILP model for 3D TAM optimization Problem $P^{H}_{DTSV,||}$

Objective:

Minimize $\sum_{i=1}^{|M|} u_i$

Subject to the Constraints:

$t_{max} = \max_{i=1}^{|M|} t_i$

$c_i \geq x_{ij} \cdot t_j \quad \forall i, j = i..|M|$

$u_i \geq 0$

$u_i - t_{max} \cdot y_i \leq 0$

$u_i - c_i \leq 0$

$c_i - u_i + t_{max} \cdot y_i \leq t_{max}$

$\sum_{j=1}^{|M|} x_{ij} \cdot w_j \leq W_{max} \quad \forall i$

$w_i + p_{i-1} \cdot w_{i-1} + p_{i+1} \cdot w_{i+1} \leq W_{max} \quad \forall i \in die, i \neq |M|$

$x_{ii} = 1 \quad \forall i$

$x_{ij} = x_{ji} \quad \forall i, j$

$1 - x_{ij} \geq x_{ik} - x_{jk} \geq x_{ij} - 1 \quad \forall i \neq j \neq k$

$x_{i,i+1} \leq p_{i+1} \quad \forall i \in die, i \neq |M|$

$x_{i+2,i+1} \leq p_{i+1} \quad \forall i \in die, i \neq |M|$

$\sum_{i=2}^{|M|} d_i \leq TSVmax_i$

$d_i \geq \sum_{j=k}^{|M|} w_j \cdot x_{kj} \quad \forall i \in die, i \neq |M|, k = i..|M|$

$d_i \geq w_j \quad \forall i \in die, i \neq |M|, j = i..|M|$

$d_i \geq w_j + p_{j-1} \cdot w_{j-1} + p_{j+1} \cdot w_{j+1} \quad \forall i \in die, j \geq i$

$y_1 = 1$

$y_i \geq \frac{1}{1-i} \sum_{j=1}^{i-1} (x_{ij} - 1) - C \quad \forall i > 1$

$p_i = 1 \quad \forall i \in die$

$p_{i+1} \leq w_i - w_{i+1} \quad \forall i \in die, i \neq |M|$

$p_{i+1} \leq w_{i+2} - w_{i+1} \quad \forall i \in die, i \neq |M|$

As before, to determine the test architecture and test time for a given die (SOC) with a given TAM width, the TAM design method in [18] for daisychain TestRail architectures [32] was utilized. For problem instances with hard dies, the test times (cycles) and TAM widths for different dies are listed in Table 8.6. Note that test pins were assigned to dies based on their sizes in order to avoid very large test times for any individual die.

For a fixed value of TSV_{max} and range of values of W_{max}, Table 8.7 presents results for P^{H}_{MTS} and P^{S}_{MTS} for the two benchmark SICs. They are compared against optimized results for only the final post-bond stack test (referred to as P^{H}_{FS} and P^{S}_{FS}). The ILP models were run and optimal results obtained using XPRESS-MP [34]. The CPU times for the experiments on an AMD Opteron 250 with four gigabytes of memory were in the range of a few seconds to 8 min. In this table, Column 1 lists the optimization model that was run and Column 2 lists the benchmark SIC. Column 3 shows the maximum number of TSVs allowed for each die (TSV_{max}), while Column 4 lists the number of available test pins W_{max} at the lowest die. Column 5 represents the total test length (cycles) for all four post-bond test insertions in a five-die stack. Columns 6 gives the percent reduction in test time over the optimization case for only the final stack test. Let the test length for P^{S}_{FS} be T_1 and let the test length for P^{S}_{MTS} be T_2. Then, the percentage reduction is $((T_1 - T_2)/T_1) \cdot 100\,\%$. The calculation for hard dies is performed in the same

Fig. 8.33 ILP model for 3D TAM optimization Problem $P^{S}_{DTSV,||}$

Objective:

Minimize $\sum_{i=1}^{N} u_i$

Subject to the Constraints:

$$
\begin{array}{l}
t_{max} = \max_{i=1}^{|M|} t_i \\
c_i \geq v_{ij} \qquad \forall i, j = i..|M| \\
v_{ij} \geq 0 \qquad \forall i, j \\
v_{ij} - t_{max} \cdot x_{ij} \leq 0 \qquad \forall i, j = 1..|M| \\
-\sum_{n=1}^{k_i} (g_{jn} \cdot t_j(n)) + v_{ij} \leq 0 \qquad \forall i, j \\
\sum_{n=1}^{k_i} (g_{jn} \cdot t_j(n)) - v_{ij} + t_{max} \cdot x_{ij} \leq t_{max} \qquad \forall i, j \\
u_i \geq 0 \\
u_i - t_{max} \cdot y_i \leq 0 \\
u_i - c_i \leq 0 \\
c_i - u_i + t_{max} \cdot y_i \leq t_{max} \\
z_{ijk} \geq 0 \\
z_{ijk} - t_{max} \cdot x_{jk} \leq 0 \\
-wi + z_{ijk} \leq 0 \\
w_i - z_{ijk} + t_{max} \cdot x_{jk} \leq t_{max} \\
\sum_{i=2}^{|M|} d_i \leq TSV_{max} \\
\sum_{i=2}^{|M|} d_i \leq TSVmax_i \\
d_i \geq \sum_{j=k}^{|M|} z_{jkj} \qquad \forall i \in die, i \neq |M|, k = i..|M| \\
d_i \geq w_j \qquad \forall i \in die, i \neq |M|, j = i..|M| \\
d_i \geq w_j + pw_{j-1} + pw_{j+1} \qquad \forall i \in die, j \geq i \\
\sum_{j=1}^{|M|} z_{jij} \leq W_{max} \qquad \forall i \\
w_i + pw_{i-1} + pw_{i+1} \leq W_{max} \qquad \forall i \in die, i \neq |M| \\
x_{ii} = 1 \qquad \forall i \\
x_{ij} = x_{ji} \qquad \forall i, j \\
1 - x_{ij} \geq x_{ik} - x_{jk} \geq x_{ij} - 1 \qquad \forall i \neq j \neq k \\
x_{i,i+1} \leq p_{i+1} \qquad \forall i \in die, i \neq |M| \\
x_{i+2,i+1} \leq p_{i+1} \qquad \forall i \in die, i \neq |M| \\
y_1 = 1 \\
y_i \geq \frac{1}{1-i} \sum_{j=1}^{i-1} (x_{ij} - 1) - M \qquad \forall i > 1 \\
p_i = 1 \qquad \forall i \in die \\
p_{i+1} \leq w_i - w_{i+1} \qquad \forall i \in die, i \neq |M| \\
p_{i+1} \leq w_{i+2} - w_{i+1} \qquad \forall i \in die, i \neq |M| \\
pw_i \geq 0 \qquad \forall i \\
pw_i - W_{max} \cdot p_i \leq 0 \qquad \forall i \\
pw_i - w_i \leq 0 \qquad \forall i \\
w_i - pw_i + W_{max} \cdot p_i \leq W_{max} \qquad \forall i
\end{array}
$$

manner. The percentage reductions for P^{H}_{FS} and P^{S}_{FS} are zero by default. Columns 7 through 10 show the resulting test schedule for the 3D-SIC at each test insertion, where the symbol "‖" indicates parallel testing of dies, and a "," represents serial testing. Finally, Column 11 gives the number of test pins needed for each die. From Table 8.7 it can be seen that the proposed method can provide considerable reductions in test time over the optimization methods of Sect. 8.2 that are targeted toward final stack test.

For a different number of TSVs (TSV_{max}), Fig. 8.34a and b show the variation in test time T with an increase in the number of test pins W_{max} for SIC 1 and SIC 2.

Table 8.6 Test lengths and number of test pins for hard dies used in optimization

Die name	d695	f2126	p22810	p34392	p93791
Test length	106391	700665	1333098	2743317	2608870
Test pins	15	20	25	25	30

From the figures, it can be seen that both TSV_{max} and W_{max} determine which dies should be tested in parallel and thus the total test time for the stack. For a given value of TSV_{max}, increasing W_{max} does not always decrease the test time for P^{H}_{MTS}, although test time never increases. Similarly, increasing TSV_{max} for a given W_{max} does not always decrease the test time. These Pareto-optimal points can easily be seen in Fig. 8.34 for both benchmarks. Furthermore, it can be seen that optimizing for the final stack test does not always reduce test time when multiple test insertions are considered. In fact, it is often the case that the test time is higher when increasing the values of TSV_{max} and W_{max}.

For optimization of 3D-SICs with soft dies, Pareto-optimality is almost non-existent when W_{max} is varied; see Fig. 8.35. This is due to the fact that when dies in the stack are soft, it is almost always possible to find one die for which adding an extra test pin reduces the overall test time. When test time stops decreasing, the TSV limits for the problem instance have been reached and no further optimization is possible. Once again, as expected, the optimization method for multiple test insertions outperforms previous models.

It should be noted that for all optimizations with the same values of TSV_{max} and W_{max}, the stack configuration (SIC 2) with the largest die at the highest layer and the smallest die at the lowest layer is the best for reducing test time while using the minimum number of TSVs. This is because the most complex dies with the longest test times are tested in the fewest insertions. For example, the die at the top of the five-die stack is only tested once, while the die at the bottom is tested four times.

For $P^{H}_{DTSV,||}$, two extra test pins were added to the hard die for each EXTEST TAM on a die. For the highest and lowest dies in the stack, this implies an addition of two test pins, while for other dies it implies an addition of four test pins. This value was chosen because it is the smallest addition of test pins necessary to result in reduced EXTEST times. It is possible to perform EXTEST with only two additional pins added to a die, but this results in unreasonably long EXTEST times. For the problem instances it is assumed, without loss of generality, that each die has 10,000 functional TSVs, requiring 20 patterns. It has been reported that the number of tests for TSVs is likely to grow logarithmically with TSV count [42]. In order to make comparisons between $P^{H}_{DTSV,--}$ and $P^{H}_{DTSV,||}$, either two or four test pins were added to the total TAM width of the hard dies. As can be seen in Fig. 8.36, there is a large dependence of test length on both TSV_{max} and W_{max}. These parameters and the hard TAM widths determine whether serial testing or parallel testing of TSVs with the dies is adopted to lower test times. For a majority of values for TSV_{max}

Table 8.7 Experimental results and comparisons for multiple test insertions

Optimization framework	SIC	TSV_{max}	W_{max}	Test length (cycles)	Reduction (%)	Test schedule 1 (Die 1-2)	Test schedule 2 (Die 1-2-3)	Test schedule 3 (Die 1-2-3-4)	Test schedule 4 (Die 1-2-3-4-5)	No. of test pins used per Die
P_{FS}^{H}	SIC 1	70	49	21254500	0.00	1,2	1,2‖3	1,2‖3,4	1‖5,2‖3,4	30,25,25,20,15
	SIC 1	70	50	19091000	0.00	1,2	1,2,3	1,2,3‖4	1,2‖5,3‖4	30,25,25,20,15
	SIC 1	70	69	10819000	0.00	1‖2	1‖2‖3	1‖2‖3,4	1‖2‖3,4‖5	30,25,25,20,15
	SIC 1	70	70	10819000	0.00	1‖2	1‖2‖3	1‖2‖3,4	1‖2‖3,4‖5	30,25,25,20,15
P_{MTS}^{H}	SIC 1	70	49	12901800	39.30	1‖2	1,2‖3	1‖4,2‖3	1‖4,2‖3,5	30,25,25,20,15
	SIC 1	70	50	11042700	42.16	1‖2	1,2‖3	1‖2,3‖4	1,2‖5,3‖4	30,25,25,20,15
	SIC 1	70	69	9554160	11.69	1‖2	1‖2‖3	1‖3‖4,2	1‖2‖3,4‖5	30,25,25,20,15
	SIC 1	70	70	8959880	17.18	1‖2	1‖2‖3	1,2‖3‖4	1‖4‖5,2‖3	30,25,25,20,15
P_{FS}^{S}	SIC 2	50	49	13366500	0.00	1‖2	1‖2‖3	1‖2‖3‖4	1‖2‖3‖4,5	22,16,6,4,28
	SIC 2	50	50	12149400	0.00	1‖2	1‖2‖3	1‖2‖3‖4	1‖2‖3‖4,5	24,16,6,4,30
	SIC 2	50	51	12149400	0.00	1‖2	1‖2‖3	1‖2‖3‖4	1‖2‖3‖4,5	24,17,6,4,30
P_{MTS}^{S}	SIC 2	50	49	9636860	27.90	1‖2	1‖2‖3	1‖2,3‖4	1‖2‖3,4‖5	3,18,28,20,29
	SIC 2	50	50	9467830	22.07	1‖2	1‖2‖3	1,2‖3‖4	1‖2,3‖4‖5	22,18,10,16,24
	SIC 2	50	51	9392830	22.69	1‖2	1‖2‖3	1‖2‖3‖4	1‖2,3‖4‖5	3,22,10,16,24

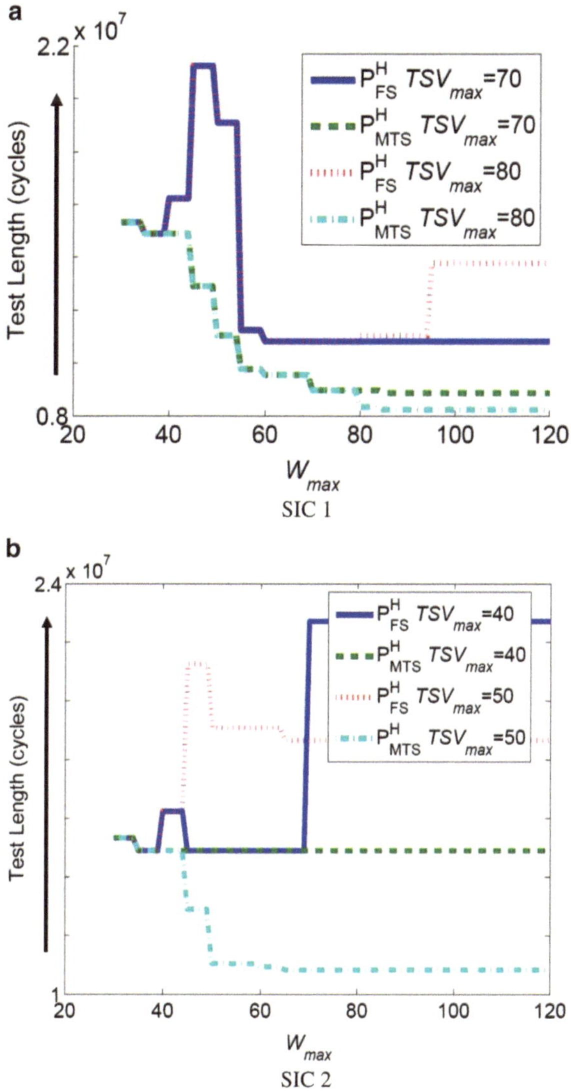

Fig. 8.34 The test time with respect to TSV_{max} and W_{max} for SIC 1 and SIC 2 with hard dies and multiple test insertions (**a**) SIC 1 (**b**) SIC 2

and W_{max} for the TAM widths chosen, parallel testing leads to lower test times. Note here that these optimizations apply only to the final stack test. As such, SIC 2 results in longer test times than SIC 1, as expected from Sect. 8.2.

In $P^S_{DTSV,--}$ and $P^S_{DTSV,||}$, there is more opportunity for test-time optimization and the results differ from those obtained with hard dies, as shown in Fig. 8.37. Because full control is available over both die-internal and die-external TAM widths, trade-offs for parallel testing are magnified. Test pins that are not utilized for die-internal tests and instead used to support die-external tests tend to result in longer global test times. This is due to the complexity of the modules tested during

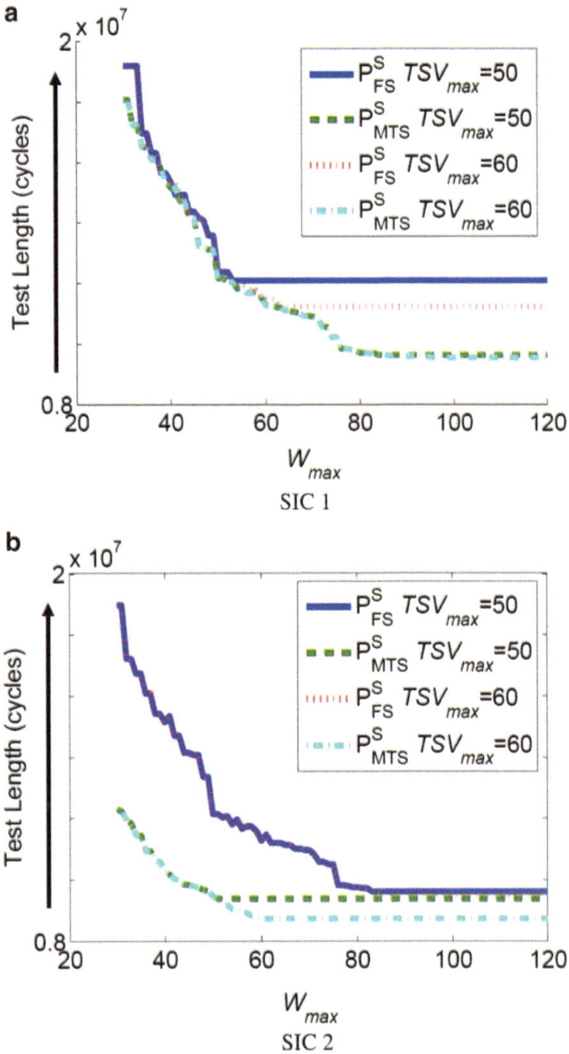

Fig. 8.35 The test time with respect to TSV_{max} and W_{max} for SIC 1 and SIC 2 with soft dies and multiple test insertions (**a**) SIC 1 (**b**) SIC 2

die-internal tests when compared to TSV tests. Many more patterns are needed for die-internal tests, and as such, test pins devoted to these tests reduce test time more than test pins dedicated to die-external tests. Because it requires at least four test pins to result in test-time reduction for a single die-external test, this utilization of test pins does not provide much benefit. Therefore, thin wrappers and serial testing of TSVs result in lower test times than fat wrappers and parallel testing for 3D stacks with soft dies, as seen in Fig. 8.37.

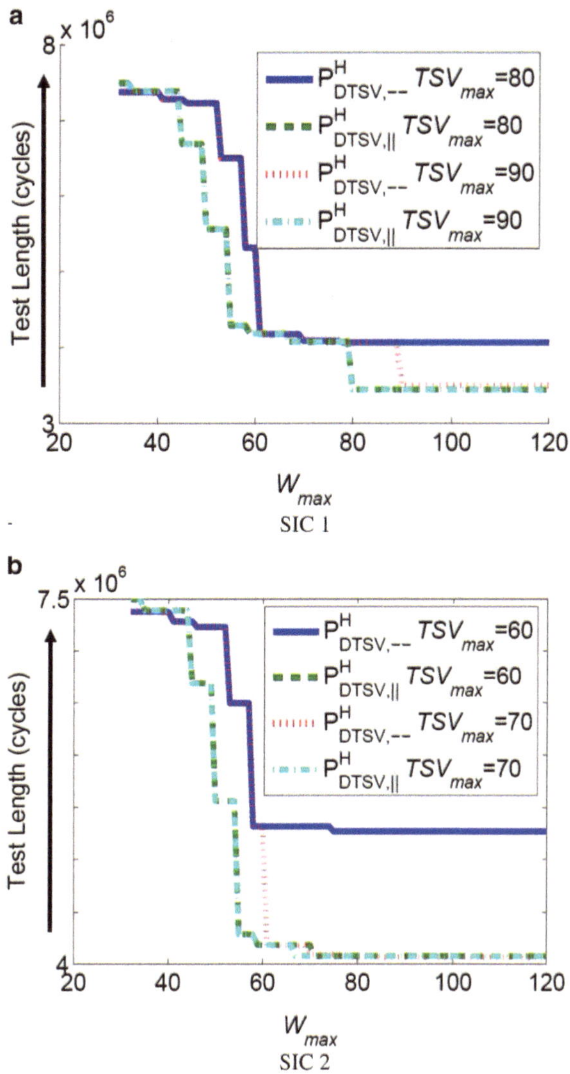

Fig. 8.36 The test time with respect to TSV_{max} and W_{max} for SIC 1 and SIC 2 with hard dies including die-external tests (**a**) SIC 1 (**b**) SIC 2

8.5.1 Summary

Section Summary

- The test architecture and schedule optimization problem has been extended to consider any/all desired post-bond test insertions, post-bond die EXTEST, and dedicated test TSV limits between adjacent dies.
- If multiple test insertions are required, optimizing for only the final stack test tends to result in significantly sub-optimal test architectures.

Fig. 8.37 The test time with respect to TSV_{max} and W_{max} for SIC 1 and SIC 2 with soft dies including die-external tests (**a**) SIC 1 (**b**) SIC 2

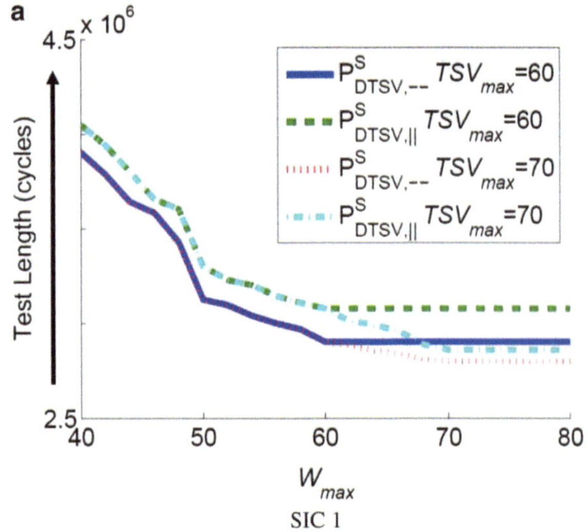

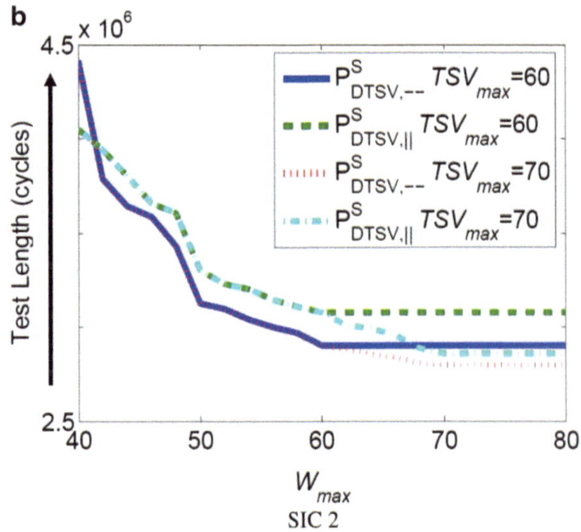

- For soft dies, standard ("thin") wrapper design results in lower test times when compared to "fat" wrappers.

8.6 Conclusions

This chapter has dealt extensively with post-bond test architecture and test scheduling problems in the context of an ILP mathematical model and optimization solution. ILP solutions were presented for three different, but realistic, 3D test-architecture optimization problems. These solutions produce optimal TAM designs and test schedules under test-access constraints for three design scenarios, referred to as hard, soft, and firm dies. The optimization models are rigorously derived and their optimality is formally established. It is shown that these methods result in significantly decreased test time over heuristic methods for 3D TAM design.

The effects of a number of parameters on the optimization results were examined; the results provide insights to designers on how to design for testability prior to 3D integration, how to design the 3D stack itself, and how to best allocate the limited test resources for the 3D SIC. The parameters of interest include the placement of dies in a stack, the availability of test TSVs and test pins. It has been demonstrated that, given test bandwidth and TSV constraints, pareto-optimality exists in the solution space. This finding is especially significant for the hard die case, although it is also present in the case of firm dies. These results imply that, using the derived models, designers can explore many test solutions for the hard and firm die scenarios in order to prevent sub-optimal allocation of test resources. Firm and soft dies are shown to provide lower test times with an increase in test resources. Because both scenarios result in lower test times than the case of hard dies, designers can consider 2D and 3D TAM design as related to optimization problems.

Generalized and rigorous optimization methods to minimize test time for multiple test insertions were further derived. These methods also provide optimal solutions for several additional problem instances of interest, e.g., final stack test only and testing of any subset of partial stacks. This extended optimization model is defined to be globally optimal in the case of multiple test insertions. Furthermore, optimization techniques were derived to consider die-external and die-internal tests during test-time minimization. Results show that optimization methods that only consider the final stack test provide significantly suboptimal results (higher test times) if multiple test insertions are carried out. Optimizations for hard dies that take into account die-internal and die-external tests show that in general, fat wrappers reduce test time more than thin wrappers. This is not the case with soft dies, because the availability of fewer test pins for die-internal tests tends to lead to more test time.

Chapter 9
Conclusions

This book has covered a wide array of topics relating to the design-for-test and test optimization for 3D SICs. These topics were chosen to explore test in many scenarios, including memory-on-memory, memory-on-logic, and logic-on-logic stacks. BIST and probing techniques were explored for pre-bond TSV and scan test. Methods for yield assurance, including BISR architectures and wafer matching, were explained in detail. Optimizations for reducing test cost were also covered, including flows to reduce the delay overhead of DfT architectures and to optimize TAM architectures and test schedules to reduce test time. Together, the topics covered by this book offer an extensive and in-depth look at the cutting-edge of 3D test for students, teachers, researchers, and industry practitioners.

Chapter 2 examined methods for yield assurance in the production of 3D SICs. The first half of the chapter explored different approaches to wafer matching and analyzed the impact of matching choices on yield and cost. These choices include the type (running or static) and size of wafer repositories, various matching algorithms and criteria, and more. The second half of the chapter covered yield assurance for memory stacks. Models were derived for unique resistive faults that can arise in 3D stacks that use TSVs as bitlines and wordlines. Various redundancy architectures were presented, including redundancy sharing between dies, global redundancy architectures, and the use of redundant layers. The implications of these redundancy architectures combined with wafer matching, as they impact manufacturing cost and stack yield, were explored.

Chapter 3 discussed BIST architectures for pre-bond testing of TSVs. Electrical models were derived for a variety of defects that can occur in the TSV pillar or insulating sheath. Several BIST architectures were examined for the pre-bond testing of the TSV pillars. These methods ranged from RAM-like read and write operations on the TSV pillar to ring oscillators. The strengths and limitations of each architecture were thoroughly discussed.

Chapter 4 offers an alternative to BIST for pre-bond TSV test through probing. The present state-of-the-art in probe card design is provided, including a discussion of probing limitations. A combination of on-die and pin electronic architectures was explained to enable the probing of TSVs based on emerging standards and

B. Noia and K. Chakrabarty, *Design-for-Test and Test Optimization Techniques for TSV-based 3D Stacked ICs*, DOI 10.1007/978-3-319-02378-6_9,

probe limitations. This was achieved by shorting together multiple TSVs with a single probe needle. Detailed results and analysis were provided to demonstrate the feasibility, accuracy, limitations, and cost of pre-bond TSV probing. Furthermore, an optimization technique, which reduced test time by allowing for the accurate test of multiple TSVs shorted through the same probe needle simultaneously, was discussed.

Chapter 5 extended the probing paradigm presented in Chap. 4 to enable both pre-bond TSV and scan test through the same probing paradigm. This was achieved by utilizing TSVs as test inputs and outputs along with reconfigurable scan chains. These scan chains were switched between pre-bond test modes utilizing TSVs and post-bond test modes that utilize the emerging die wrapper standard. An analysis of the feasibility, speed, and cost of the method was provided.

Chapter 6 explored a retiming-based test architecture optimization for reducing the impact of the test architectures described in Chaps. 4 and 5 on the post-bond functional speed of the stack. Performance improvements were achieved by moving slack in the 3D circuit to offset the additional delay caused by test-related boundary registers at the interface between TSVs and die logic. A new retiming algorithm was developed to allow the limited movement of logic between dies and enable the decomposition of complex logic gates. The effectiveness, benefits, and drawbacks of retiming at both the stack- and die-level were detailed.

Chapter 7 explained the emerging standards for 3D SICs important for the present and future of 3D SIC design. These standards include the P1838 die wrapper for die and stack testing and the JEDEC Wide I/O standard for memory-on-logic stacks. The implementation of these standards was discussed in detail, including the test implications of and test solutions for the emerging standards.

Finally, Chap. 8 developed an optimization method for the 3D TAM and test schedule to minimize the test time for post-bond stacks. A detailed and flexible ILP model was developed that enabled test optimization for any or all post-bond test insertions, for varying test pin and test TSV constraints, for external tests, and more. A detailed analysis of the test time reduction achievable through the optimization, including comparisons to other algorithms, was provided.

In summary, this book provides a detailed and thought-provoking examination of the state-of-the-art techniques in 3D SIC test. For students and teachers it provides a complete and in-depth overview of a wide variety of test-related topics. For researchers and industry practitioners, it provides detailed descriptions and analysis of cutting-edge solutions for immediate implementation or as a starting point for new research directions. The authors sincerely hope that this book has fulfilled the need for the first comprehensive book on 3D IC test, and that readers may now go on to make further contributions to the exciting field of 3D testing.

References

1. M. L. Bushnell and V. D. Agrawal, "Essentials of Electronic Testing for Digital, Memory, and Mixed-Signal VLSI Circuits", Boston; Kluwer Academic Publishers, 2000.
2. L.-T. Wang, C.-W. Wu, and X. Wen. **VLSI Test Principles and Architectures: Design for Testability**. Elsevier Inc. , San Francisco, USA, 2006.
3. P. Garrou, C. Bower, and P. Ramm. **Handbook of 3D Integration—Technology and Applications of 3D Integrated Circuits**. Wiley-VCH, Weinheim, Germany, August 2008.
4. Yuan Xie, "Processor Architecture Design Using 3D Integration Technology", *VLSI Design*, pp. 446–451, 2010.
5. S. Wong, A. El-Gamal, P. Griffin, Y. Nishi, F. Pease, and J. Plummer, "Monolithic 3D Integrated Circuits", *VLSI Technology, Systems and Applications*, pp. 1–4, 23–25, 2007.
6. J. Feng, Y. Liu, P. Griffin, and J. Plummer, "Integration of Germanium-on-Insulator and Silicon MOSFETs on a Silicon Substrate", *Electronic Device Letters*, pp. 911–913, 2006.
7. F. Crnogorac, D. Witte, Q. Xia, B. Rajendran, D. Pickard, Z. Liu, A. Mehta, S. Sharma, A. Yasseri, T. Kamins, S. Chou, and R. Pease, "Nano-Graphoepitaxy of Semiconductors for 3D Integration", *Microelectronic Engineering*, pp. 891–894, 2007.
8. Tezzaron Octopus. http://www.tezzaron.com/memory/Octopus.html Accessed March 2011.
9. "Samsung Develops 8GB DDR3 DRAM Using 3D Chip Stacking Technology." http://www.samsunghub.com/2010/12/07/samsung-develops-8gb-ddr3-dram-using-3d-chip-stacking-technology/ Accessed March 2011.
10. U. Kang et al. "8Gb 3D DDR3 DRAM Using Through-Silicon-Via Technology", In *Proc. International Solid State Circuits Conference (ISSCC)*, pp. 130–132, February 2009.
11. S. Das, A. Chandrakasan, and R. Reif, "Design Tools for 3-D Integrated Circuits," In *Proc. IEEE Asia South Pacific Design Automation Conference (ASP-DAC)*, pp. 53–56, 2003.
12. B. Noia, K. Chakrabarty, and Y. Xie, "Test-Wrapper Optimization for Embedded Cores in TSV-Based Three-Dimensional SOCs", *Proc. IEEE International Conference on Computer Design*, pp. 70–77, 2009.
13. Physorg.com http://pda.physorg.com/_news170952515.html
14. K. Chakrabarty, "Optimal Test Access Architectures for System-on-a-Chip", *ACM Transactions on Design Automation of Electronic Systems*, vol. 6, pp. 26–49, January 2001.
15. M.L. Flottes, J. Pouget, and B. Rouzeyre, "Sessionless Test Scheme: Power-Constrained Test Scheduling for System-on-a-Chip", *Proc. of the 11th IFIP on VLSI-SoC*, pp. 105–110, 2001.
16. E.G. Coffman, Jr., M.R. Garey, D.S. Johnson and R.E. Tarjan. "Performance Bounds for Level-Oriented Two-Dimensional Packing Algorithms", *SIAM J. Computing*, vol. 9, pp. 809–826, 1980.
17. X. Dong and Y. Xie. "System-level Cost Analysis and Design Exploration for 3D ICs", *Proc. of Asia-South Pacific Design Automation Conference (ASP-DAC)*, pp. 234–241, Jan. 2009.

B. Noia and K. Chakrabarty, *Design-for-Test and Test Optimization Techniques for TSV-based 3D Stacked ICs*, DOI 10.1007/978-3-319-02378-6,

18. S.K. Goel and E.J. Marinissen, "Control-Aware Test Architecture Design for Modular SOC Testing", *European Test Workshop*, pp. 57–62, 2003.
19. E.J. Marinissen, V. Iyengar, and K. Chakrabarty. "A Set of Benchmarks for Modular Testing of SOCs", In *International Test Conference*, pp. 519–528, Oct. 2002.
20. *IEEE Std. 1500: IEEE Standard Testability Method for Embedded Core-Based Integrated Circuits. IEEE Press*, New York, 2005.
21. M.R. Garey and D.S. Johnson, "Computers and Intractability—A Guide to the Theory of NP-Completeness", New York: Freeman, 1979.
22. V. Iyengar, K. Chakrabarty, and E. J. Marinissen, "Test Wrapper and Test Access Mechanism Co-optimization for System-on-Chip", *JETTA*, vol. 18, pp. 213–230, 2002.
23. S.K. Goel, E.J. Marinissen, A. Sehgal and K. Chakrabarty, "Testing of SOCs with Hierarchical Cores: Common Fallacies, Test-Access Optimization, and Test Scheduling", *IEEE Transactions on Computers*, vol. 58, pp. 409–423, March 2009.
24. E. Larsson, K. Arvidsson, H. Fujiwara, and Z. Peng, "Efficient Test Solutions for Core-based Designs," *TCAD*, vol. 23, no. 5, pp. 758–775, 2004.
25. W. R. Davis et al., "Demystifying 3D ICs: the Pros and Cons of Going Vertical," *IEEE Design and Test of Computers*, vol. 22, no. 6, pp. 498–510, 2005.
26. P.-Y. Chen, W. C.-W. Wu, and D.-M. Kwai, "On-Chip Testing of Blind and Open-Sleeve TSVs for 3D IC Before Bonding", *Proc. IEEE VLSI Test Symposium*, pp. 263–268, 2010.
27. D. Lewis and H.-H. Lee, "A Scan-Island Based Design Enabling Pre-bond Testability in Die-Stacked Microprocessors", *Proc. International Test Conference*, pp. 1–8, 2007.
28. Y. Xie, G. H. Loh, and K. Bernstein, "Design Space Exploration for 3D Architectures," J. *Emerg. Technol. Comput. Syst.*, 2(2):65–103, 2006.
29. G. Loh, Y. Xie, and B. Black, "Processor Design in 3D Die Stacking Technologies," *IEEE Micro* Vol 27, No. 3, pp. 31–48, 2007.
30. E. J. Marinissen, S. K. Goel, and M. Lousberg, "Wrapper Design for Embedded Core Test," *Proc. Int'l Test Conf.*, pp. 911–920, 2000.
31. Y. Huang et al., "Optimal Core Wrapper Width Selection and SOC Test Scheduling Based on 3-D Bin Packing Algorithm,"' In *International Test Conference*, pp. 74–82, 2002.
32. S.K. Goel and E.J. Marinissen. "SOC Test Architecture Design for Efficient Utilization of Test Bandwidth,"' *ACM Transactions on Design Automation of Electronic Systems*, 8(4):399–429, 2003.
33. Q. Xu and N. Nicolici, "Resource-Constrained System-on-a-Chip Test: A Survey", *IEE Proc.: Computers and Digital Techniques*. vol. 152, pp. 67–81, Jan. 2005.
34. FICO. Xpress-MP. http://www.fico.com/en/Products/DMTools/Pages/FICO-Xpress-Optimization-Suite.aspx
35. E.J. Marinissen, J. Verbree, and M. Konijnenburg, "A Structured and Scalable Test Access Architecture for TSV-Based 3D Stacked ICs", *VLSI Test Symposium*, 2010.
36. B. Noia, S.K. Goel, K. Chakrabarty, E.J. Marinissen, and J. Verbree, "Test-Architecture Optimization for TSV-Based 3D Stacked ICs", *European Test Symposium*, 2010.
37. B. Noia, K. Chakrabarty and E. J. Marinissen, "Optimization Methods for Post-bond Die-Internal/External Testing in 3D Stacked ICs", *Proc. IEEE International Test Conference*, 2010.
38. H. Chen, J.-Y. Shih, S.-W. Li, H.-C. Lin, M.-J. Wang, C.-N. Peng. "Electrical Tests for Three-Dimensional ICs (3DICs) with TSVs.", *International Test Conference 3D-Test Workshop*, 2010.
39. L. Jiang, Q. Xu, K. Chakrabarty, and T.M. Mak, "Layout-Driven Test-Architecture Design and Optimization for 3D SoCs under Pre-Bond Test-Pin-Count Constraint", *International Conference on Computer-Aided Design*, pp. 191–196, 2009.
40. U. Kang et. al., "8 Gb 3-D DDR3 DRAM Using Through-Silicon-Via Technology", *IEEE Solid-State Circuits*, vol. 45, no. 1, pp. 111–119, 2010.
41. M. Cho, C. Liu, D. Kim, S. Lim, and S. Mukhopadhyay, "Design Method and Test Structure to Characterize and Repair TSV Defect-Induced Signal Degradation in 3D System", *Proc. IEEE Conference on Computer-Aided Design*, pp. 694–697, 2010.

42. E.J. Marinissen and Y. Zorian, "Testing 3D Chips Containing Through-Silicon Vias", *International Test Conference*, E 1.1, 2009.
43. H.-H.S. Lee and K. Chakrabarty, "Test Challenges for 3D Integrated Circuits", *IEEE Design & Test of Computers*, vol. 26, pp. 26–35, September/October 2009.
44. K. Puttaswamy and G. H. Loh, "The Impact of 3-Dimensional Integration on the Design of Arithmetic Units," in *IEEE International Symposium on Circuits and Systems*, 2006.
45. K. Puttaswamy and G. H. Loh, "Thermal Herding: Microarchitecture Techniques for Controlling Hotspots in High-Performance 3D-Integrated Processors," *IEEE High Performance Computer Architecture*, pp. 193–204, 2007.
46. C.C. Liu, I. Ganusov, M. Burtscher, and S. Tiwari, "Bridging the Processor-Memory Performance Gap with 3D IC Technology," *IEEE Design & Test of Computers*, 22(6):556–564, November/December 2005.
47. B. Black, D.W. Nelson, C. Webb, and N. Samra, "3D Processing Technology and its Impact on iA32 Microprocessors", *In Proc. International Conference on Computer Design (ICCD)*, pp. 316–318, October 2004.
48. X. Wu, Y. Chen K. Chakrabarty, and Y. Xie, "Test-Access Mechanism Optimization for Core-Based Three-dimensional SOCs", *IEEE International Conference on Computer Design*, pp. 212–218, 2008.
49. L. Jiang, L. Huang, and Q. Xu, "Test Architecture Design and Optimization for Three-Dimensional SoCs", *Design, Automation, and Test in Europe*, pp. 220–225, 2009.
50. K. Lee, "Trends in Test", Keynote talk presented at *IEEE Asian Test Symposium*, December 2010.
51. P. Holmberg, "Automatic Balancing of Linear AC Bridge Circuits for Capacitive Sensor Elements", *Instrumentation and Measurement Technology Conference*, pp. 475–478, 2010.
52. Qmax. QT2256 - 320 PXI. http://www.qmaxtest.com/in/Automated%20Test%20Equipment/qt2256pxifea.html Accessed January 2011.
53. T. Yasafuku, K. Ishida, S. Miyamoto, H. Nakai, M. Takamiya, T. Sakurai, and K. Takeuchi, "Effect of Resistance of TSV's on Performance of Boost Converter for Low-Power 3D SSD with NAND Flash Memories", *3D System Integration*, pp. 1–4, 2009.
54. J. Broz and R. Rincon, "Probe Needle Wear and Contact Resistance", talk presented at *Southwestern Test Workshop*, June 1998.
55. K. Kataoka, S. Kawamura, T. Itoh, T. Suga, K. Ishikawa, and H. Honma, "Low Contact-Force and Compliant MEMS Probe Card Utilizing Fritting Contact", *IEEE Conference on Micro Electro Mechanical Systems*, pp. 364–367, 2002.
56. 45nm PTM LP Model. http://ptm.asu.edu/modelcard/LP/45nm_LP.pm Accessed January 2011.
57. J. Leung, M. Zargari, B.A. Wooley, and S.S. Wong, "Active Substrate Membrane Probe Card", *Electron Devices Meeting*, pp. 709–712, 1995.
58. D. Williams and T. Miers, "A Coplanar Probe to Microstrip Transition", *IEEE Transactions on Microwave Theory and Techniques*, pp. 1219–1223, 1988.
59. O. Weeden, "Probe Card Tutorial", http://www.accuprobe.com/Downloads/Probe%20Card%20Tutorial.pdf Accessed January 2010.
60. L.-R. Huang, S.-Y. Huang, S. Sunter, K.-H. Tsai, and W.-T. Cheng, "Oscillation-Based Prebond TSV Test", *IEEE Transactions on Computer-Aided Design*, vol.32, no.9, pp. 1440–1444, 2013.
61. S. Deutsch and K. Chakrabarty, "Non-Invasive Pre-Bond TSV Test Using Ring Oscillators and Multiple Voltage Levels", *Proc. Design, Automation, and Test Conference in Europe*, pp. 18–22, 2013.
62. T. Thorolfsson et al., "Design Automation for a 3DIC FFT Processor for Synthetic Aperture Radar: A Case Study", *Proc. IEEE Design Automation Conference*, 2009.
63. J. Verbree, E. J. Marinissen, P. Roussel, and D. Velenis, "On the Cost-Effectiveness of Matching Repositories of Pre-Tested Wafers for Wafer-to-Wafer 3D Chip Stacking", *Proc. IEEE European Test Symposium*, pp. 36–41, 2010.
64. J. Schat, "Fault Clustering in Deep-Submicron CMOS Processes", *Proc. Design, Automation, and Test in Europe*, pp. 511–514, 2008.

65. S. Reda, G. Smith, And L. Smith, “Maximizing the Functional Yield of Wafer-to-Wafer 3-D Integration”, *IEEE Transactions on VLSI Systems*, vol. 17, pp. 1357–1362, 2009.
66. M. Taouil, S. Hamdioui, J. Verbree, and E. J. Marinissen, “On Maximizing the Compound Yield for 3D Wafer-to-Wafer Stacked ICs”, *Proc. IEEE International Test Conference*, pp. 1–10, 2010.
67. L. Jiang, Y. Liu, L. Duan, Y. Xie and Q. Xu, “Modeling TSV Open Defects in 3D-Stacked DRAM”, *Proc. IEEE International Test Conference*, pp. 1–9, 2010.
68. G. Loh, “3D-Stacked Memory Architectures for Multi-Core Processors”, *Proc. International Symposium on Computer Architecture*, pp. 453–464, 2008.
69. M. F. Hilbert. High-Density Memory Utilizing Multiplexers to Reduce Bitline Pinch Constraints. *US Patent 6,377,504*, 2002.
70. Tezzaron Semiconductor. http://www.tezzaron.com/ Accessed September 2013.
71. M. Taouil and S. Hamdioui, “Layer Redundancy Based Yield Improvement for 3D Wafer-to-Wafer Stacked Memories”, *Proc. IEEE European Test Symposium*, pp. 45–50, 2011.
72. K. Puttaswamy et al., “3D-Integrated SRAM Components for High-Performance Microprocessors”, *IEEE Transactions on Computers*, vol. 58, pp. 1369–1381, 2009.
73. L. Jiang, R. Ye, and Q. Xu, “Yield Enhancement for 3D-Stacked Memory by Redundancy Sharing Across Dies”, *Proc. IEEE/ACM International Conference on Computer-Aided Design*, pp. 230–234, 2010.
74. X. Wang, D. Vasudevan, and H.-H. S. Lee, “Global Built-In Self-Repair for 3D Memories with Redundancy Sharing and Parallel Testing”, *Proc. IEEE 3D Systems Integration Conference*, pp. 1–8, 2012.
75. S.-K. Lu, C.-L. Yang, Y.-C. Hsiao, and C.-W. Wu “Efficient BISR Techniques for Embedded Memories Considering Cluster Faults”, *IEEE Transactions on Very Large Scale Integration Systems*, vol. 18, pp. 184–193, 2010.
76. K. Smith et al., “Evaluation of TSV and Micro-Bump Probing for Wide I/O Testing”, *Proc. IEEE International Test Conference*, pp. 1–10, 2011.
77. O. Yaglioglu, and N. Eldridge, “Direct Connection and Testing of TSV and Microbump Devices Using NanoPierce^{TM} Contactor for 3D-IC Integration” *Proc. IEEE VLSI Test Symposium*, pp. 96–101, 2012.
78. B. Leslie and F. Matta, “Wafer-Level Testing with a Membrane Probe”, *IEEE Design and Test of Computers*, pp. 10–17, 1989.
79. Y. Zhang, Y. Zhang, and R. B. Marcus, “Thermally Actuated Microprobes for a New Wafer Probe Card”, *Microelectromechanical Systems*, vol. 8, pp. 43–49, 1999.
80. Y.-W. Yi, Y. Kondoh, K. Ihara, and M. Saitoh, “A Micro Active Probe Device Compatible with SOI-CMOS Technologies”, *Microelectromechanical Systems*, vol. 6, pp. 242–248, 1997.
81. W. Zhang, P. Limaye, R. Agarwal, and P. Soussan, “Surface Planarization of Cu/Sn Microbump and its Application in Fine Pitch Cu/Sn Solid State Diffusion Bonding”, *Proc. Conf. on Electronics Packaging Technology*, pp. 143–146, 2010.
82. Y. Ohara et al., “10 μm Fine Pitch Cu/Sn Micro-Bumps for 3-D Super-Chip Stack,” *Proc. 3D System Integration*, pp. 1–6, 2009.
83. G. Katti et al., “Temperature Dependent Electrical Characteristics of Through-Si-Via (TSV) Interconnections,” *International Interconnect Technology Conference*, pp. 1–3, 2010.
84. D. Velenis, E. J. Marinissen, and E. Beyne, “Cost Effectiveness of 3D Integration Options,” *Proc. 3D Systems Integration Conference*, pp. 1–6, 2010.
85. Open core circuits. http://www.opencores.org Accessed March 2012.
86. D.H. Kim, K. Athikulwongse, and S.K. Lim, “A Study of Through-Silicon-Via Impact on the 3D Stacked IC Layout”, *Proc. IEEE International Conference on Computer-Aided Design*, 2009.
87. A.D. Trigg et al., “Design and Fabrication of a Reliability Test Chip for 3D-TSV,” *Proc. Electronic Components and Technology Conference*, pp. 79–83, 2010.
88. S.L. Wright, P. S. Andry, E. Sprogis, B. Dang, and R. J. Polastre, "Reliability Testing of Through-Silicon Vias for High-Current 3D Applications," *Proc. Electronic Components and Technology Conference*, pp. 879–883, 2008.

89. L.J. Thomas, H. K. Kow, and S. Palasundram, “Current Capacity Evaluation of a Cantilever Probe,” *Proc. Electronic Manufacturing Technology Symposium*, pp. 1–6, 4–6, 2008.
90. K.M. Butler, J. Saxena, A. Jain, T. Fryars, J. Lewis, and G. Hetherington, “Minimizing Power Consumption in Scan Testing: Pattern Generation and DFT Techniques,” *Proc. IEEE International Test Conference*, pp. 355–364, 2004.
91. X. Wen, “Towards the Next Generation of Low-Power Test Technologies,” *Proc. IEEE International Conference on ASIC*, pp. 232–235, 2011.
92. D. Gizopoulos, K. Roy, P. Girard, N. Nicolici, and X. Wen, “Power-Aware Testing and Test Strategies for Low-Power Devices,” *Proc. Design, Automation and Test in Europe*, pp. 10–14, 2008.
93. B. Noia and K. Chakrabarty, “Pre-bond Probing of TSVs in 3D Stacked ICs”, *Proc. IEEE International Test Conference*, pp. 1–10, 2011.
94. B. Noia, S. Panth, K. Chakrabarty and S. K. Lim, “Scan Test of Die Logic in 3D ICs Using TSV Probing”, accepted for publication in *Proc. IEEE International Test Conference*, pp. 1–8, 2012.
95. C.E. Leiserson and J.B. Saxe, “Optimizing Synchronous Systems”, *Journal of VLSI Computing Systems*, vol. 1, pp. 41–67, 1983.
96. G. De Micheli, “Synchronous Logic Synthesis: Algorithms for Cycletime Minimization”, *IEEE Trans. Computer-Aided Design*, vol. 10, pp. 63–73, 1991.
97. J. Monteiro, S. Devadas, and A. Ghosh, “Retiming Sequential Circuits for Low Power”, *Proc. IEEE International Conference on CAD*, pp. 398–402, 1993.
98. S. Dey and S. Chakradhar, “Retiming Sequential Circuits to Enhance Testability”, *Proc. IEEE VLSI Test Symposium*, pp. 28–33, 1994.
99. T.C. Tien, H.P. Su, and Y.W. Tsay, “Integrating Logic Retiming and Register Placement”, *Proc. IEEE Intl. Conf. Computer-Aided Design*, pp. 136–139, 1998.
100. O. Sinanoglu and V. D. Agrawal, “Eliminating the Timing Penalty of Scan”, *Journal of Electronic Testing: Theory and Applications*, vol. 29, Issue 1, pp. 103–114, 2013.
101. Nangate Open Cell Library. http://www.si2.org/openeda.si2.org/projects/nangatelib Accessed March 2012.
102. Kenneth P. Parker. *The Boundary Scan Handbook*. Springer-Verlag, 2003.
103. IEEE 3D-Test Working Group. http://grouper.ieee.org/groups/3Dtest/ Accessed August 2013.
104. Wide I/O Single Data Rate (JEDEC Standard JESD229). JEDEC Solid State Techonology Association. http://www.jedec.org// Accessed August 2013.
105. E.J. Marinissen, C.-C. Chi, J. Verbree, and M. Konijnenburg, “3D DfT Architecture for Pre-Bond and Post-Bond Testing”, *Proc. IEEE 3D Systems Integration Conference*, pp. 1–8, 2010.
106. F. Brglez, D. Bryan, and K. Kozminski, “Combinational Profiles of Sequential Benchmark Circuits”, *Proc. IEEE International Symposium on Circuits and Systems*, pp. 1924–1934, 1989.
107. S.K. Goel et al., “Test Infrastructure Design for the Nexperia^{TM} Home Platform PNX8550 System Chip”, *Proc. IEEE Design, Automation, and Test in Europe*, pp. 108–113, 2004.
108. S. Deutsch et al., “DfT Architecture and ATPG for Interconnect Tests of JEDEC Wide-I/O Memory-on-Logic Die Stacks”, *Proc. IEEE International Test Conference*, pp. 1–10, 2012.

MIX
Papier aus verantwortungsvollen Quellen
Paper from responsible sources
FSC® C105338

If you have any concerns about our products,
you can contact us on
ProductSafety@springernature.com

In case Publisher is established outside the EU,
the EU authorized representative is:
Springer Nature Customer Service Center GmbH
Europaplatz 3, 69115 Heidelberg, Germany

Printed by Libri Plureos GmbH
in Hamburg, Germany